Linear Integrated Circuit Applications Using Electronics Workbench®

Hardware and Simulation Exercises

Borris
St. Clair County Community College

Prentice Hall
Upper Saddle River, New Jersey   Columbus, Ohio

Cover art: Interactive Image Technologies, Ltd.
Editor: Scott Sambucci
Production Editor: Rex Davidson
Design Coordinator: Karrie Converse-Jones
Cover Designer: Rod Harris
Production Manager: Patricia A. Tonneman
Marketing Manager: Ben Leonard

This book was printed and bound by Victor Graphics. The cover was printed by Victor Graphics.

© 2000 by Prentice-Hall, Inc.
Pearson Education
Upper Saddle River, New Jersey 07458

All rights reserved. No part of this book may be reproduced, in any form or by any means, without permission in writing from the publisher.

The representation of screens from Electronics Workbench used by permission of Interactive Images Technologies Ltd. Copyright 1993, 1995, 1997, 1998 by Interactive Images Technologies Ltd. Unauthorized use, copying, or duplicating is strictly prohibited.

Printed in the United States of America

10 9 8 7 6 5 4 3 2 1

ISBN: 0-13-280835-8

Prentice-Hall International (UK) Limited, *London*
Prentice-Hall of Australia Pty. Limited, *Sydney*
Prentice-Hall Canada Inc., *Toronto*
Prentice-Hall Hispanoamericana, S. A., *Mexico*
Prentice-Hall of India Private Limited, *New Delhi*
Prentice-Hall of Japan, Inc., *Tokyo*
Prentice-Hall (Singapore) Pte. Ltd., *Singapore*
Editora Prentice-Hall do Brasil, Ltda., *Rio de Janeiro*

PREFACE

Most instructors of electronics are aware of the importance of hands-on hardware applications of electronics. We all know this is essential and is an indispensable part of electronic engineering technology. This book gives the student hardware applications as well as simulation exercises using Electronics Workbench® software. Although this book is primarily about linear integrated circuit applications, discrete devices and some digital integrated circuits are included.

The instructor that likes hardware applications will be pleased to know that hardwired projects are included in the chapters; especially in Chapter 10, which is dedicated mostly to hardware projects. For the instructor that wishes to take advantage of the new simulation techniques that most progressive industries now use, there are numerous laboratory exercises.

The purpose of this book is to emphasize areas of electronics that industry wants current and future employees to understand. Employers want employees that know how to think on their feet, develop a plan of work, and follow through by completing the company project. This means that we must teach some theory, show students the steps in planning and having good documentation, and have them complete quality application projects that have meaning in the real world. It also means we must let go of how we used to do it and try to stay in the mainstream of the rapidly changing technological world.

Linear Integrated Circuit Applications Using Electronics Workbench®: with Hardware and Simulation Exercises provides a discussion of the key concepts of the exercises including design examples and formulas needed to analyze circuit operation. All exercises provide an opportunity to analyze the schematic, take voltage readings, conduct waveform analysis, and modify the circuit to illustrate the importance of knowing how and why the circuit works. I recommend the Prentice Hall text: *Operational Amplifiers & Linear Integrated Circuits*, by Coughlin, as an accompaniment to this book. At the end of each chapter are questions and problems, suggestions on hardware projects, and key concepts. The instructor of this book should emphasize that students read the Key Concepts section prior to beginning the laboratory exercises. This section includes theory, suggested formulas, and design examples.

A list of all components used for the exercises is contained in Appendix D. Data sheets from manufacturers are shown in Appendix A. Components used in these exercises are easily obtainable from mail order electronic supply companies. A list of companies that offer components either individually or as a complete package for the course is in Appendix C.

Important Information for Students

Please read the Key Concepts at the end of the chapter prior to attempting the laboratory exercise. At most, 15 to 20 minutes of preparation time reading the Key Concepts will result in saved time. If you are a newcomer to Electronics Workbench you will want to do the short exercise in Appendix B. This book assumes that you have completed DC and AC, discrete semiconductor devices, and have a hands-on knowledge of the typical electronics laboratory equipment.

When reading the Key Concepts you may want to jot down the formulas on a sheet of paper and later transfer the formulas to the margin beside the steps in the lab exercise. Remember that formulas are put in the form that the author of the book prefers, but in reality different forms are essentially the same. Your instructor will tell you what formula format he or she prefers.

It must be emphasized that not everything is given to you for all the exercises. In some cases you will need to consult another textbook or a data sheet in the appendix. Remember, this is not a cookbook, but a guide for you so that you may use your thinking and creative skills. Computer analysis of the circuits allows you to analyze each of the circuits. It also allows you to change values without worrying whether components will be damaged. Creative construction of circuits based on your knowledge of theory is one of the most valuable parts of the simulation software. Hardwired circuits provide a challenge to you in that you learn best component layout and are able to make measurements with equipment. The software you are using for this course has been in use for ten years and is available as a student package for about the cost of a textbook.

Electronics Workbench registered software is available from Interactive Image Technologies by calling 1(800)263-5552. The software is based on proven Spice Circuit Analysis and will work with any circuit in this book when used properly.

Acknowledgments

I would like to thank my wife Autumn for her patience while I spent many hours in front of the computer at home. I also would like to thank my secretary Trish Schultz for putting the manuscript into a nice clean format. My students have been an integral part of this project also as they have done the experiments and made suggestions when things did not go quite right. I would also like to thank an octogenarian friend of mine, Orville Beeler, whose interest in electronics and computers continues to inspire me to learn.

I also thank the editors who have supported me when illness twice interrupted the completion of this manual. I also thank my colleagues Matt Morabito and Paul Johnson for their ideas and support. I hope this meets their need for more hands-on applications.

ANDERSONIAN LIBRARY
WITHDRAWN FROM LIBRARY STOCK
UNIVERSITY OF STRATHCLYDE

TABLE OF CONTENTS

CHAPTER 1
Name ______________________
INTRODUCTION TO OPERATIONAL AMPLIFIERS

INTRODUCTION

Operational amplifiers were originally used to perform mathematical operations. In the following exercises you will connect the "op-amp" to perform a variety of operations. For example, in this chapter of simulation exercises you will see how the op-amp performs without the feedback loop closed. You will see how zero crossing detectors work, and how positive and negative voltage level detectors perform. Most importantly, you will see some practical applications of voltage level detection by examining a level detector that sounds a buzzer when sound level at the input creates a voltage that exceeds a reference voltage. Then you will modify and redesign the sound detector to create a burglar alarm circuit that will sound an alarm when a door entry is detected.

OBJECTIVES

- Learn how to identify leads on a 741A integrated circuit (IC) chip.
- Understand the input output characteristics of the op-amp.
- Learn how to connect the inverting and non-inverting input terminals to observe zero voltage crossing in the open loop condition.
- Connect the op-amp to observe positive and negative input voltage levels.
- Build and analyze the zero crossing detector.
- Analyze a practical application of a voltage level detector.
- Modify an application to create a new application circuit.

Exercise 1-1, The 741 Op-Amp Detecting Voltage Levels

PROCEDURE

Step 1. Load the file labeled **"ex#1-1"** from your EWB circuit files disk: **Operational Amplifier Circuit Applications Using Electronics Workbench**. Observe the package and pin layout of the 741A operational amplifier in the center of your computer screen. This is a dual in-line (DIP) package. Note the location of pin #1 if the package is rotated clockwise 90 degrees. For example, pin #1 will be on the top left when looking at the top side of the IC package. This will be true of all (DIP) integrated circuit packages. Other details are shown in Figure 1-1, as well as on your computer screen.

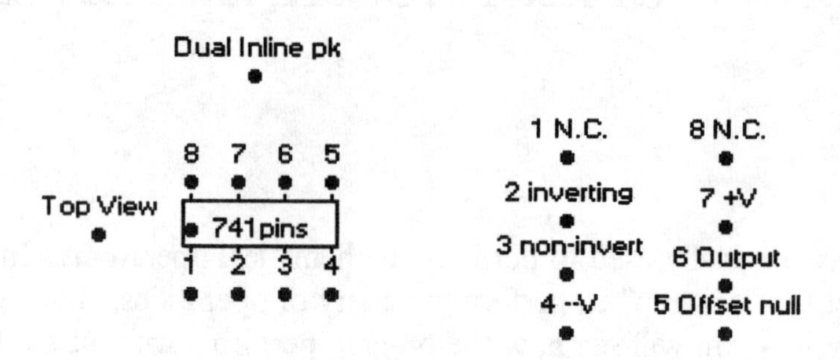

Figure 1-1

Step 2. Draw the top view of the 741 op-amp in the space below. Show the IC as if it were plugged into your real circuit hardware breadboard with pin #1 on your top left. Label the pins showing their purpose.

Step 3. Note the power supply connections on the far left of your monitor screen. The parts bin on EWB has two versions of op-amps for you to use. One (without power leads) already has a fixed plus and minus 12 volt power source connected, but not shown to you. The other has plus and minus power pins from which you can connect your desired voltages (generally ±5 volts to ±18 volts). Build the simulation circuit shown in Figure 1-2. The parts bin for the op-amp is in the menu shown by the diode symbol. Use the left mouse button to drag (roll) parts out to the breadboard space. Drag the function generator and oscilloscope out onto the breadboard. Zoom the generator and oscilloscope instrument faces by double-clicking on each instrument face. Set the function generator to 1 volt (it outputs in peak value) peak and 1 kilohertz. Connect the function generator and oscilloscope as shown. Draw the resulting waveforms (on the oscilloscope face in Figure 1-3) when the function generator is set for triangle wave output.

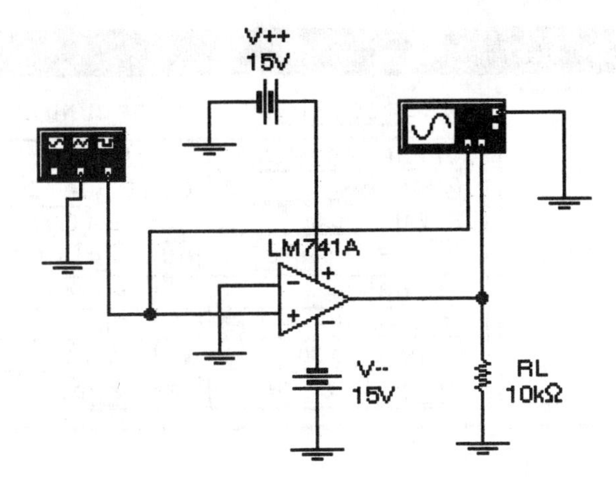

Figure 1-2

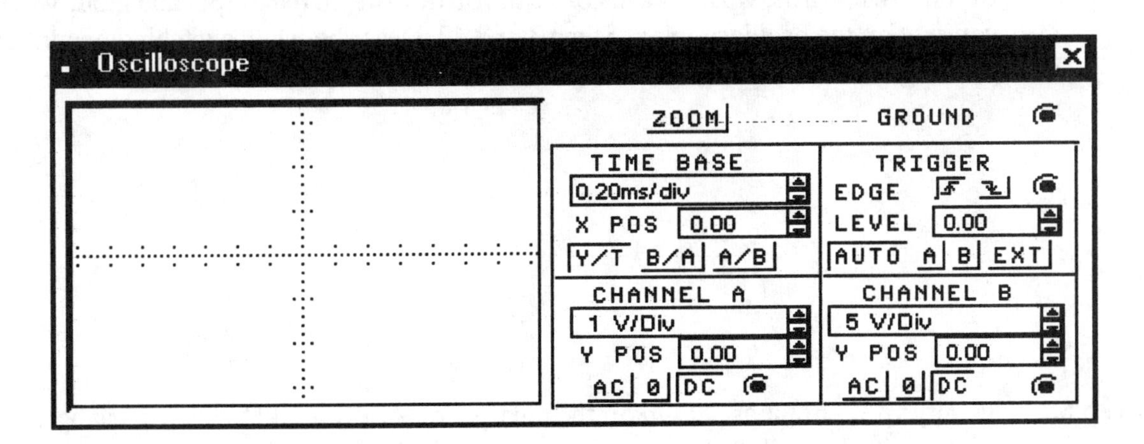

Figure 1-3

Step 4. Repeat the process in Step 3 above using a sine wave input. Record on the oscilloscope face in Figure 1-4.

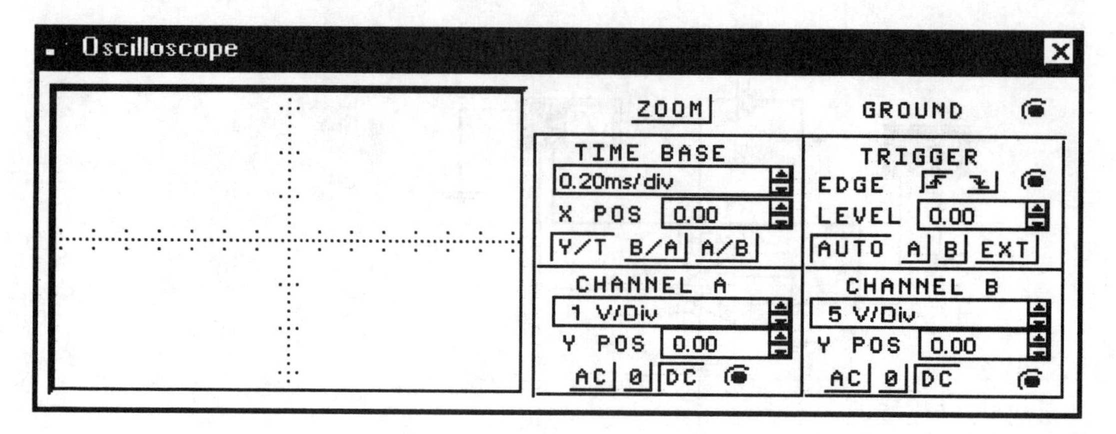

Figure 1-4

Step 5. In your own words, what conclusions can you draw from the output and input waveforms you saw on the oscilloscope in Steps 3 and 4? Describe when switching occurs on both crossings and where saturation occurs.

Step 6. Modify your circuit as shown on the next page. Apply a 1 kHz, 2v p-p triangle wave to this circuit. Describe when switching occurs on both crossings.

Step 7. For Step 6 above, draw the waveforms on the oscilloscope face for the triangle wave input in Figure 1-6.

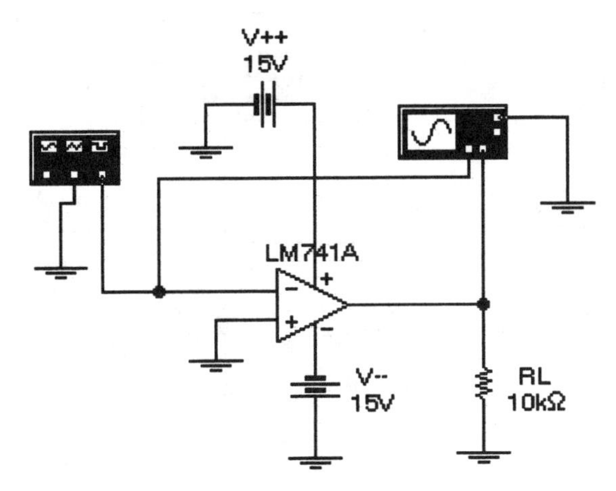

Figure 1-5

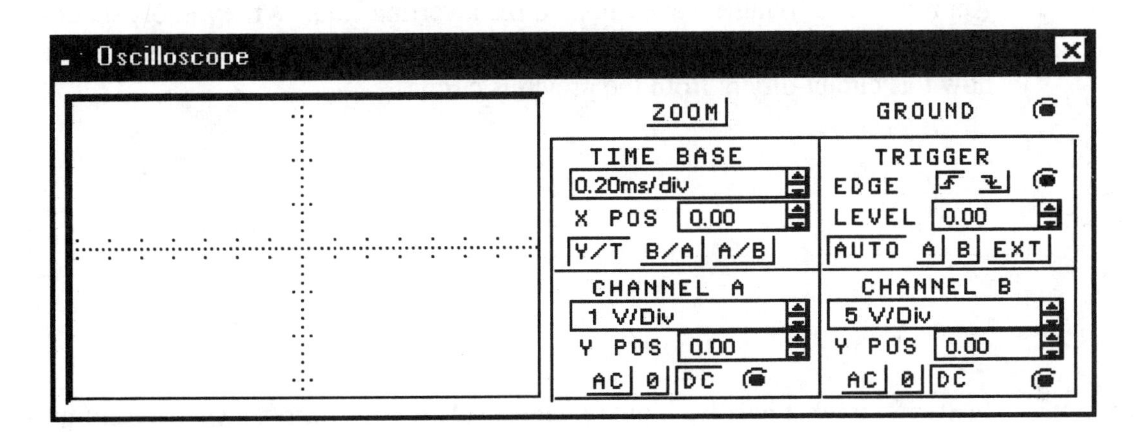

Figure 1-6

Step 8. Modify the circuit in Step 6 to detect a positive-going voltage. You will need to connect a positive reference voltage to the non-inverting input. Connect the circuit in Figure 1-7.

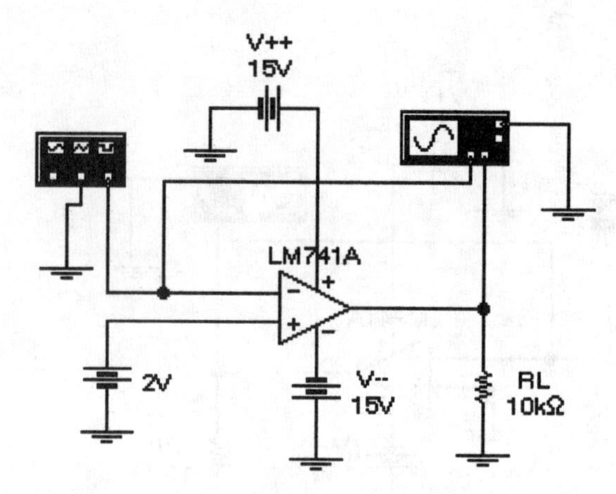

Figure 1-7

Apply a 8v p-p triangle wave signal to the inverting lead. Write a short statement on the operation of this circuit. In this statement, show (using waveform results) and describe how this circuit differs from the previous circuit.

Step 9. In Step 8, at what positive input voltage level does the output switch to saturation? Does the output switch to positive saturation or negative saturation?

__________ V

+Vsat or −Vsat __________

Step 10. Reverse the reference battery in the non-inverting lead of the same circuit above, Figure 1-7. At what voltage does the switching occur now? Does the output switch to positive or negative saturation?

__________ V __________

Step 11. ***Optional.*** Revise the previous circuit, Figure 1-7, and place the battery in the inverting lead. Apply the signal to the non-inverting input. Explain the output versus input waveform.

When operating an op-amp without feedback the voltage gain is very high; for example, it is approximately 200,000 for the 741 operational amplifier IC.

Key Concepts
Operational Amplifiers

A high open loop gain means that a small voltage difference between the input leads results in the output voltage going to a positive or negative saturation level.

Saturation voltage in a practical op-amp is normally 1 to 2 volts less than the V++ or V-- power supply voltage values.

When the non-inverting input (+) is above (more positive) the inverting input (-) the output will go to the positive saturation voltage level. This output voltage will be about 1 volt less than V++.

When the inverting input (-) is above (more positive) the non-inverting input (+) the output will go to the negative saturation level. This output voltage will be about 1 volt less than V--.

In a non-inverting zero crossing detector, when Vout makes a positive-going transition from -Vsat to +Vsat , it shows that the input signal crossed zero in the positive direction.

In an inverting zero crossing detector when Vout makes a positive going transition from -Vsat to +Vsat, it shows that the input signal crossed zero in the negative direction.

A positive voltage level detector operates as a comparator. A positive reference is placed on one input; for example, the inverting input. The input signal is applied to the non-inverting input. When the input signal goes more positive than the positive reference voltage the output voltage goes to +Vsat. When the input signal goes below the reference voltage then the output voltage goes to -Vsat.

A negative voltage level detector operates in a similar manner. A negative reference voltage is placed on one of the inputs; for example, the inverting input. The input signal is applied to the non-inverting input. When the input signal goes below the reference voltage the output voltage goes to the -Vsat level.

Exercise 1-2, Applications of Level Detection

INTRODUCTION

A voltage level detector has many applications in the real world. Sound levels, gas and liquid levels, forces, mechanical stresses, moisture content, light level, and temperature are some of the analog values that sometimes need to be sensed. The devices that sense these changes are called transducers. The circuit below shows a typical application. The transducer that is used depends on the analog values that need to be sensed.

The circuit below, Figure 1-8, comes from the textbook **OPERATIONAL AMPLIFIERS & LINEAR INTEGRATED CIRCUITS** by Coughlin, R.F. and Driscoll, F.F., 5th ed., Prentice Hall.

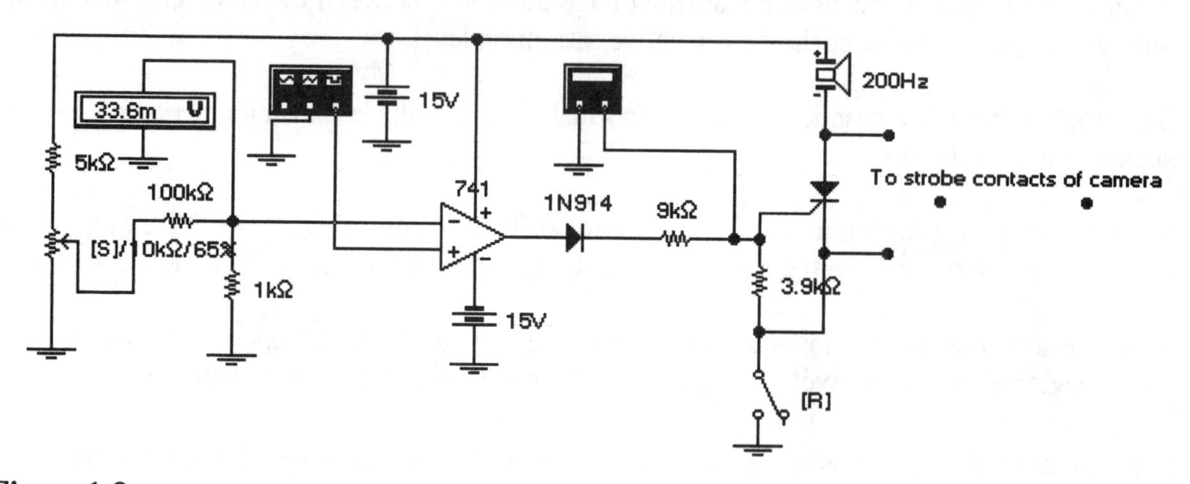

Figure 1-8

This is a sound-activated switch. Open this circuit by selecting circuit file **"ex#1-2"** from your circuits disk. The SCR is activated by a positive voltage at its gate terminal **G**. To turn the SCR on two things must happen: (1) the reset switch [R] must be closed by pressing key R, and (2) the microphone voltage must exceed the reference voltage set by switch [S]. Once the SCR turns on, it remains on until the voltage to the SCR cathode (K) is interrupted. Pressing the key [R] will disconnect the supply voltage and turn the SCR off.

PROCEDURE

Step 1. Press key **R** on your keyboard to connect the cathode of the SCR to ground. Increase the input microphone (signal generator) voltage until the alarm sounds. Record the peak voltage that activates the 200 Hz alarm. Remember that the EWB generator control sets the peak voltage of the waveform.

Vpk = _______

Step 2. Reduce the input signal voltage to a level you believe will shut the alarm off. Does the alarm shut off? _______

Activate the reset switch (highlight it with the mouse to make it an active component). Press key **R** to open the ground of the supply to the SCR. Close (remember to highlight) the **R** switch. Does the alarm sound now?

____

Step 3. Experiment with this circuit by placing the voltmeter across the anode to cathode terminals (labeled switch terminals). Measure and record the voltage from cathode to anode when the alarm is off. Measure and record the voltage from the cathode to anode when the alarm is on.

Alarm off V(a-k) = _______ Alarm on V(a-k) = _______

Note: The circuit may have to be reactivated using the on-off switch to read the correct voltage.

Step 4. Adjust the sensitivity potentiometer (key **S**) to a value above the microphone input voltage. Does this shut off the alarm? _______

Step 5. Modify the circuit so that it becomes a burglar alarm circuit. The specifications of the circuit are as follows: a normally closed switch prevents the alarm from sounding. When the switch opens, the alarm will sound; for example, when a door or a window opens. Hint: Use a logic circuit with two inputs. This circuit will replace the microphone. Set the sensitivity to the appropriate level at the other op-amp input. Draw or print out your schematic for this circuit. Explain to your instructor in your own words how your circuit is going to function without failure.

Chapter 1 Questions
Voltage Level Detectors

EWB and Hardware Laboratory Problems are Included

Fill in the correct answers to the questions below.

1. Without the feedback loop closed, a small voltage difference between the input terminals will normally cause the output voltage to go to ________________________.

2. When the non-inverting input is grounded and a positive voltage is applied to the inverting input, the output voltage of an op-amp without feedback will ________________.

3. When the non-inverting input is grounded and a negative voltage is applied to the inverting input, the output voltage of an op-amp without feedback will ________________.

4. The non-inverting input of an open feedback loop op-amp is at 1.5 volts, while an inverting input is at 1.4 volts. The voltage at its output terminal will be ____________.

5. Typical open loop gain of the 741 op-amp is ______________.

6. A zero crossing detector has a 1 volt peak-to-peak triangle waveform applied to its inverting input while its non-inverting input is grounded. When the triangle wave crosses zero in the positive direction, the output wave goes to __________ and remains there until ____.

7. An SCR may be tested by placing a short between its gate and anode while a voltage is applied as shown in Figure 1-9.

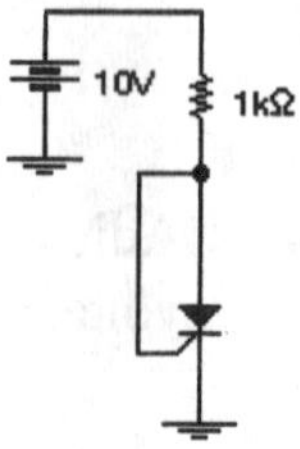

10V
1kΩ

Figure 1-9

What voltage should be measured from anode to cathode under this shorted condition? ________ V

What voltage should be measured from anode to cathode when the short is removed? ________ V

8. Obtain an SCR C106B or similar SCR from your instructor and perform this test in your hardware laboratory. Record the actual voltages you measure as follows:

   (a) Connect the circuit above without the short between anode and gate. Measure the anode to cathode voltage with your voltmeter and record.

   ____

   (b) Leave the voltmeter connected and place a short from anode to gate. What does the voltmeter read now? ____________ v

   (c) Remove the short. What does the voltmeter read? __________ v Why? Explain.

9. A voltage level detector has a reference voltage of three volts on its inverting input. The non-inverting input has a 500 Hz symmetrical triangular waveform of 10 volts peak to peak. Build this circuit on EWB. Print the schematic and its oscilloscope input and output waveforms. Take the printout to your instructor and in your own words explain the results. Build this circuit in your hands-on hardware laboratory to verify the results. Explain any differences that may occur.

10. In your hardware laboratory, build the circuit you designed on *Electronics Workbench®*. Build and test this circuit as follows:

    (a) Test the SCR part of the circuit by building the circuit below.

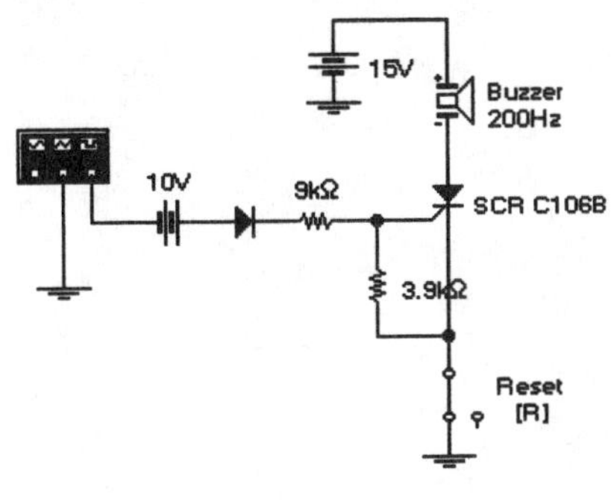

Figure 1-10

(b) Test your logic circuit to ensure that it switches from logic 0 to logic 1 (less than 0.8 volts to at least 2.5 volts).

(c) Test your voltage divider circuit so that it can provide a variable reference voltage.

(d) Connect your reference circuit to the inverting input of the op-amp. Then connect the logic circuit to the non-inverting input. Apply power to the circuit and test to see that the output voltage level of the op-amp switches properly when the logic voltage input exceeds the reference voltage input. The output of the op-amp should go to positive.

(e) If the op-amp works properly in the preceding step, then power down and connect the SCR circuit (from the diode forward only, of course). The buzzer should then sound at a frequency near your 555 astable oscillator frequency.

(f) Provide a schematic and test data for your instructor to evaluate. Prove to your instructor that you know how this circuit works by demonstrating the operation and verbally explaining why the circuit works.

CHAPTER 2

Name ______________________

BASIC OP-AMP CIRCUITS WITH FEEDBACK

INTRODUCTION

Generally, the feedback loop is closed by making a connection from the output (pin 6 of the 741) to the inverting input (pin 2). The open loop gain of the 741 op-amp is about 200,000. With the feedback connection the gain of the op-amp may be made variable from 1 to typically 100. This type of connection is called negative feedback. The gain is now called closed loop gain **A(CL)** and is independent of the open loop gain **A(OL)**. Open loop gain is still the same within the operational amplifier.

As we shall see later, the feedback loop could be made from the output back to the non-inverting input. The inverting amplifier will be connected first. In terms of theory, the inverting input pin is at virtual ground because very little current flows into the input pin. Consequently, the majority of current flows through the input resistor and continues through the feedback resistor to the output pin. Therefore, the input voltage is developed across the input resistor and the output voltage is developed across the feedback resistor. The voltage gain simply becomes the ratio of the feedback resistor (Rf) to the input resistor (R1).

The voltage gain of the non-inverting amplifier is slightly larger since the inverting pin is no longer at virtual ground; so the output voltage is developed across the feedback resistor and the resistor R1. The resulting gain is the ratio of the feedback resistor to the resistor R1, plus one. Mathematically, $(Rf / R1) + 1$. You will examine applications such as: adder, subtractor, mixer, averager, follower, and difference amplifiers.

OBJECTIVES

- Identify and draw the schematics of inverting and non-inverting amplifiers.
- Calculate and measure currents and voltages so that voltage gain and input impedance can be calculated.
- Show the output voltage waveshape time and amplitude in comparison with the time and amplitude of the input voltage wave.
- Demonstrate the phase relationship of output to input voltage.
- Design and simulate inverting and non-inverting amplifiers.
- Build a voltage follower to provide circuit isolation.
- Build and test a subtractor circuit.
- Build and test an audio mixer circuit.
- Build and test a constant current source.

Exercise 2-1, The Inverting Amplifier

PROCEDURE

Step 1. Load the file labeled "**ex #2-1**". Place ammeters in the circuit so that you measure current through each of the resistors in the circuit below, Figure 2-1. Record these currents. Does this prove to you that pin 2 is virtually at ground? ________

I (RF) = ________ I (R1) = ________ I (R2) = ________

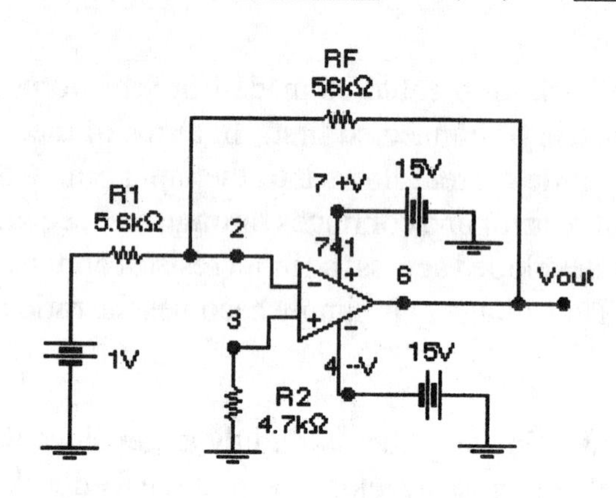

Figure 2-1

Step 2. Since most of the current that flows through RF also flows through R1, and pin 2 is at virtual ground, calculate the voltage gain of this circuit.

Av = ________

Step 3. Connect a voltmeter at the output (pin 6). From this measurement calculate the voltage gain from the formula:

Av = Vo/Vin = ________

Step 4. Replace the battery with a function generator. Set the function generator to a 1 volt peak sine wave at a frequency of 1 kHz. Measure the generator signal voltage with channel A and the output voltage with channel B. What is the voltage gain? What is the phase relationship between the input and output voltages?

Av = ________ Phase difference = ________

Step 5. Prove to yourself that the input impedance is equal to R1 by modifying your circuit as shown below, Figure 2-2. Note that if a resistance equal to R1 is connected in series with R1 the output voltage should be one-half as large as the previous measurement since the voltage Vin will be one-half of the signal voltage Vs. As a challenge to you, show your instructor the equation you have developed to prove that the input impedance is equal to R1. Write the equation down here as well.

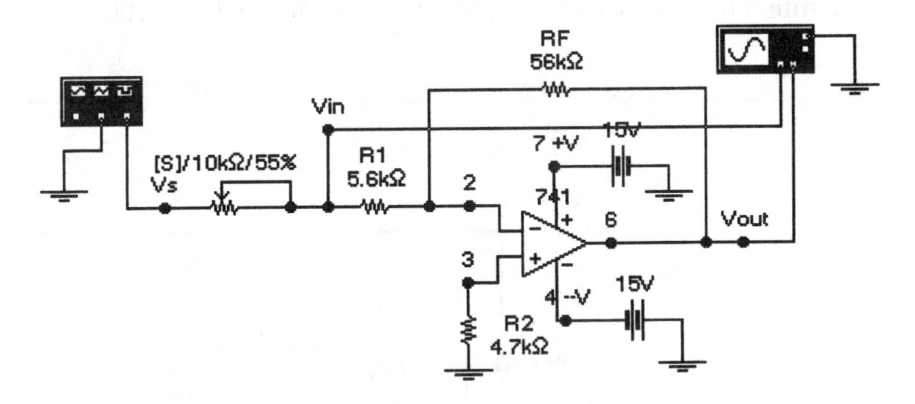

Figure 2-2

Step 6. The output versus input phase difference can be shown by placing your oscilloscope in the B/A mode. The oscilloscope will show a line from the second quadrant diagonally through the 0,0 axis down through the fourth quadrant. Try this and draw a simple sketch below. Label the voltages for both the X and Y axis on your sketch.

Exercise 2-2, Non-Inverting Amplifiers

PROCEDURE

Step 1. Load the file labeled "**ex#2-2**". Place ammeters in the circuit to measure the current through: (a) the feedback wire, (b) the output terminal to the top of the load resistor, and (c) through the 4.7 K-ohm input resistor. Record each current.

I (w) = _______ Io = _______ I (4.7k) = _______

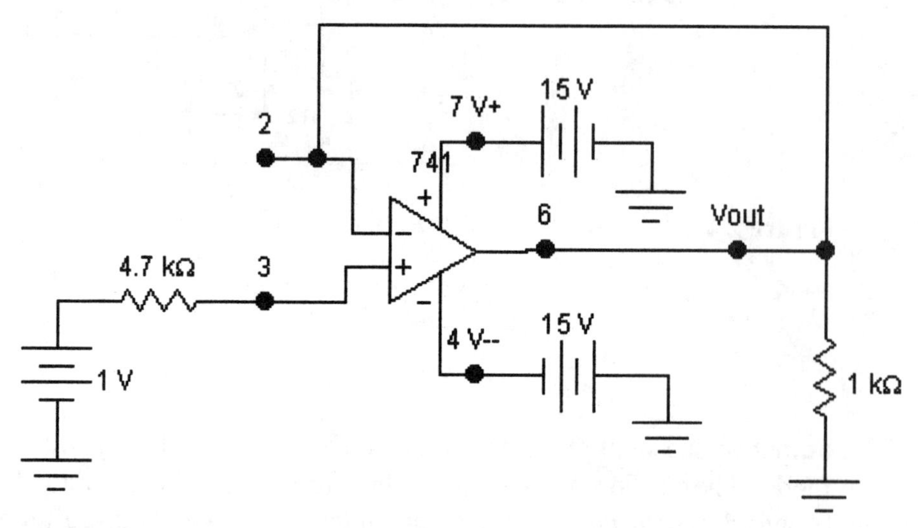

Figure 2-3

Step 2. Measure the DC voltages at pin 2 to ground and pin 3 to ground. Record these voltages. Note that the virtual ground condition does not exist.

V (2-G) = ______ V (3-G) = _______

Step 3. Measure the DC voltage across the 1 K-ohm load resistor and record. Calculate the voltage gain.

V (1K) = _______ Av = _______

Step 4. Replace the 1 volt battery with the function generator. Set the function generator to deliver a 1 volt peak sine wave at a frequency of 1 kHz. Connect channel A to the generator output and channel B across the load resistor. What is the measured voltage gain? What is the phase shift from input to output?

Av = ________ Phase shift = _________ degrees

Step 5. Place the oscilloscope in the B/A mode. Set the oscilloscope time base to 0.2 ms per division and channel A and channel B to 500 mv per division. Sketch the waveform below showing the voltage excursions in both the X and Y axis.

Step 6. Measuring the input resistance can be somewhat difficult since the input resistance can be anywhere from a couple of megohms to hundreds of megohms for some op-amps. Your measured input current and the battery voltage could be used to calculate input resistance. However, in actual practice it is difficult to measure the small amount that is present in the input. Using the series resistor method used for the inverting amplifier does not work well because of the high resistance needed. Double-clicking the left mouse button and selecting edit will allow you to get to the typical input resistance (Ri) for the 741 op-amp. Record it below.

Ri = ________

Exercise 2-3, Mathematical Operations The Adder (Summing) Circuit

PROCEDURE

Step 1. Since the inverting input of the inverting amplifier is at virtual ground and you now know that input currents to the op-amp are negligible, the current through any resistor connected to the inverting input will be the input voltage divided by the input resistor. Load the file labeled "**ex#2-3.**" Measure the current through the input resistor R1 and the feedback resistor RF and record.

I (R1) = ________ I (RF) = ________

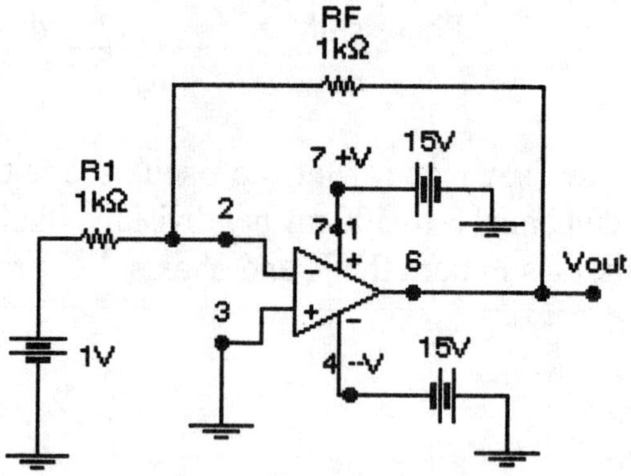

Figure 2-4

Step 2. Connect two more inputs as shown in Figure 2-5. Measure the individual input currents and the current through the feedback resistor. Show the algebraic sum of the input currents.

I (R1) = ________ I (R2) = ________

I (R3) = ________ I (RF) = ________

I (R1) + I (R2) + I (R3) = ________

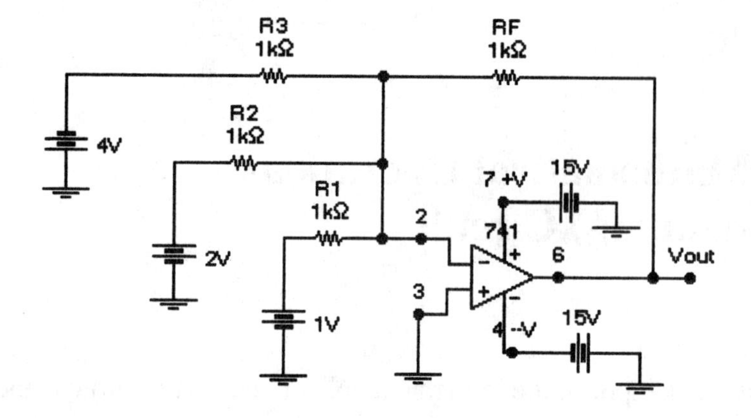

Figure 2-5

Step 3. Multiply the previously summed currents in Step 2 by the feedback resistor, RF. Record. Measure and record the voltage at the output. Are these voltages equal? If not, what might be the reason for a slight difference?

I(sum) × RF = ________ Measured Vo = ________

Step 4. Note that the output voltage is positive. You can conclude that the output of the summing amplifier is the inverted algebraic sum of the input voltages. The general expression for the summing circuit is:

$$V_O = -\left[V1\left(\frac{RF}{R1}\right) + V2\left(\frac{RF}{R2}\right) + V3\left(\frac{RF}{R3}\right)\right]$$

Figure 2-6 is a scaling circuit. The scaling circuit can solve equations such as:

2A + 3B + C where A = V1, B = V2, and C = V3

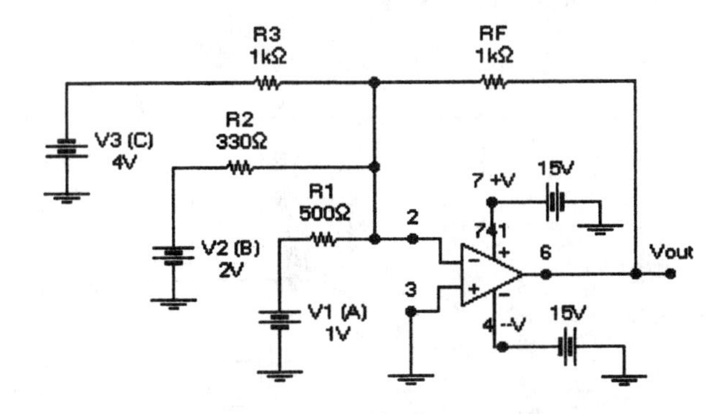

Figure 2-6

Step 5. Change the resistors R1, R2, and R3 to 1000 ohms in the circuit above, Figure 2-6. Change the resistor RF to 500 ohms. Measure the output voltage and record. Does this circuit divide? Explain.

Vo = ________ Ans. ________________________________

Step 6. Leave R1, R2, and R3 the same as in Step 5. Change RF to 2000 ohms. Measure and record the output voltage. Does this circuit multiply? Explain.

Vo = ________ Ans. ________________________________

Step 7. Leave R1, R2, and R3 the same as in Step 5. Change RF to 330 ohms. Measure the DC output voltage and record. What mathematical relationship does the absolute value of the output voltage have to the sum of the three input voltage values?

Vo = ________ Ans. ____

Step 8. Leave the circuit in Step 7 as is and place an inverter with a voltage gain of 1 (one) at its output. Measure the DC output voltage and record. What mathematical relationship does the actual output voltage have to the arithmetic sum of the three input voltages?

Vo = ________ Ans. ____

Step 9. A subtractor or "difference" amplifier finds the arithmetic difference between two inputs. The subtractor shown below acts as both an inverting and a non-inverting amplifier. The inverter has an output voltage of (−)V1 multiplied by RF/R1. The non-inverter has an output voltage of one-half V2 multiplied by [(RF/ R1) + 1]. Therefore, if all resistors are equal, the output voltage will be V2 − V1.

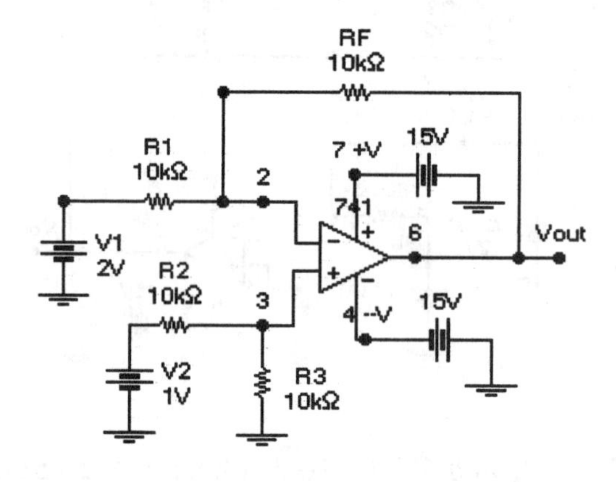

Figure 2-7

Step 10. A subtractor can be made to have whatever gain is required by changing both RF and R3 in Step 9. However, they must be changed so that RF equals R3, and R1 must remain equal to R2. Change RF and R1 to 50 K-ohm resistors. Record the output voltage. What is the gain factor compared to the output voltage in Step 9?

Vo = ________ Ans. ____

Chapter 2

Key Concepts
Op-Amps With Feedback

Closing the feedback loop will lower the voltage gain of the op-amp.

The non-inverting feedback amplifier has a voltage gain of [(RF / R1) + 1].

The non-inverting feedback amplifier has an input impedance of approximately the op-amp input impedance.

The inverting feedback amplifier has an input impedance of approximately (RF / R1).

The output signal of the inverting amplifier is 180 degrees out of phase with the input signal, while there is no phase inversion at the output of the non-inverting amplifier.

A voltage follower has a voltage gain of one. It also provides a high impedance between the load and the signal source driving the amplifier, thereby providing some isolation and decreasing the loading effect.

A subtractor will have an output voltage that is the difference between the inverting and non-inverting input signal voltages.

The differential amplifier is a subtractor where common noise at the input terminals is canceled and only the difference voltage of the input signals is amplified.

An audio mixer circuit has all inputs tied to the virtual ground input terminal (–). The inputs are therefore isolated from each other, and each input can be adjusted without affecting any other one.

A constant current source can provide a constant current to a load that fluctuates.

Chapter 2 Questions
Op-Amp Circuits with Feedback

EWB and Hardware Laboratory Problems are Included

True or False

1. ______ If both inputs of the op-amp are tied to ground, the output voltage goes to nearly zero.

2. ______ The input impedance of a voltage follower is typically one ohm.

3. ______ The output impedance of an op-amp with the feedback loop open is 100 M-ohms.

4. ______ Input signal voltage should be disconnected before the DC power is disconnected to prevent possible damage to the amplifier.

5. ______ The 741 op-amp must have a dual power supply to operate properly.

Answer the following questions for the circuit below, Figure 2-8:

6. What is the voltage gain? ________

7. What is the input impedance? ________

8. What is the approximate output impedance? ________

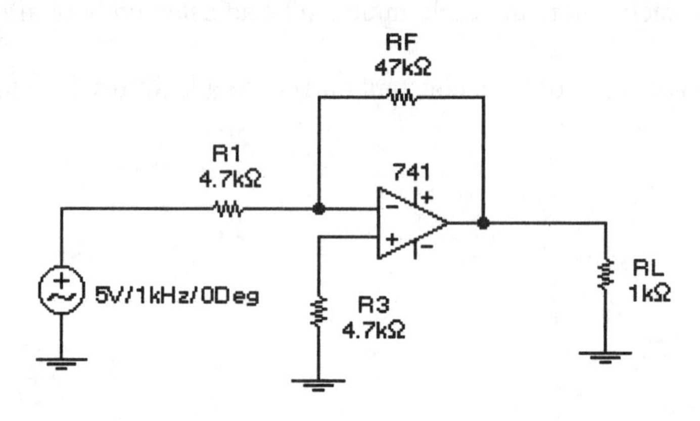

Figure 2-8

Chapter 2

9. The output is ________________________________ with the input signal.

10. Shorting the input resistor R1 would cause the input signal to ____________________ and the output signal to ______________________________.

CHAPTER 3

Name ________________

COMPARATOR CIRCUIT APPLICATIONS

INTRODUCTION

The 741 operational amplifier has a number of disadvantages when used as a comparator. Among them are: 1) the output changes relatively slowly; 2) the output voltage change is limited to +Vsat and −Vsat switching levels, therefore limiting the applications when interfacing to other devices that do not operate at those voltage levels. All comparators are susceptible to noise which may falsely trigger the comparator to an unwanted output state.

In this chapter you will use a positive feedback circuit that will reduce the possibility of false triggering. The comparator you will be using in Electronics Workbench is the LM311. There are other comparators that are much faster. You will examine the meaning of upper- and lower-level threshold voltage, hysteresis voltage, and center voltage. Also, you will look at applications such as a simple on-off controller and a controller that has adjustments for the upper and lower setpoints and the center voltage.

OBJECTIVES

- Build a zero crossing detector and show a plot of its output versus input.
- Be able to recognize from measurements the upper and lower threshold voltages.
- Determine the hysteresis voltage from your measurements of V(lowerT) and V(upT).
- Simulate a battery charger control circuit.
- Breadboard and test a battery charger controller.
- Simulate and build a window detector circuit.
- Be able to describe how a comparator operates.

Chapter 3

Exercise 3-1, Basic Comparator Concepts

PROCEDURE

Step 1. Load the file labeled "**ex#3-1.**" As shown in Figure 3-1, the comparator uses positive feedback. The key "R" controls the voltage to the inverting input. Hold the shift and the "R" key to move the wiper of the potentiometer in the opposite direction. Connect the DVM to measure the DC voltage from the wiper to ground. Connect Channel A of the oscilloscope to measure the output voltage of the op-amp. Determine the input voltages (both positive and negative) that will shift the output voltage to positive saturation and negative saturation.

V(UT) = _______ V(LT) = _______

From your measurements above, what is the hysteresis voltage?

V(H) = _______

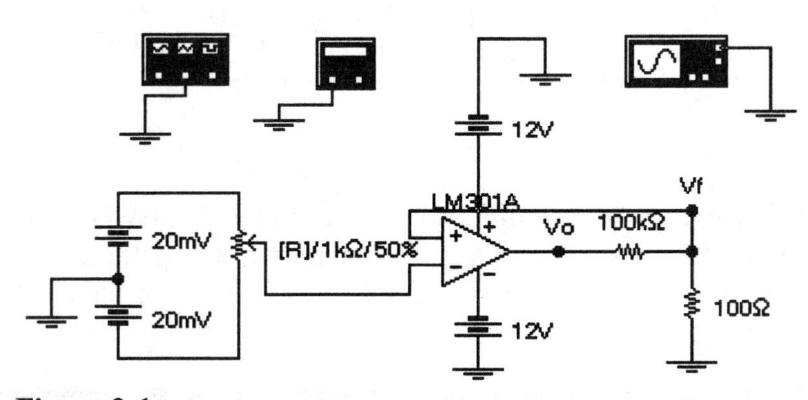

Figure 3-1

Step 2. Remove the two DC input voltage sources and replace them with the function generator. Place the function generator as shown in Figure 3-2 and set its peak amplitude to 20 millivolts of sawtooth voltage. Adjust [R] until the output begins switching between positive and negative saturation. Draw the waveform on the oscilloscope face in Figure 3-2. Record the points at which this switching occurs.

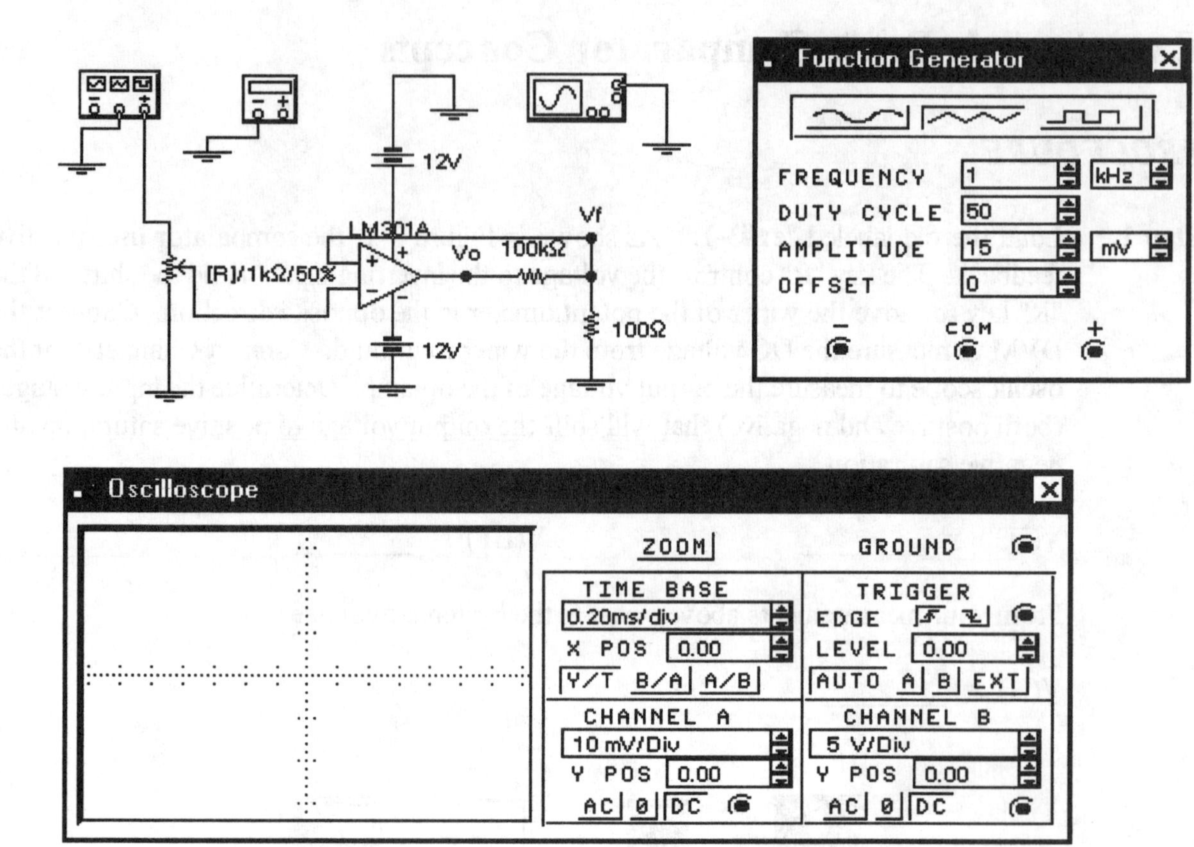

Figure 3-2

Step 3. Set the oscilloscope for **X-Y** operation. The setting on the EWB scope time-base is **B/A**. This gives a plot of Vo versus Vin. Draw the hystersis pattern and write the value of hysteresis voltage and the values of +Vsat and −Vsat on the oscilloscope face in Figure 3-3.

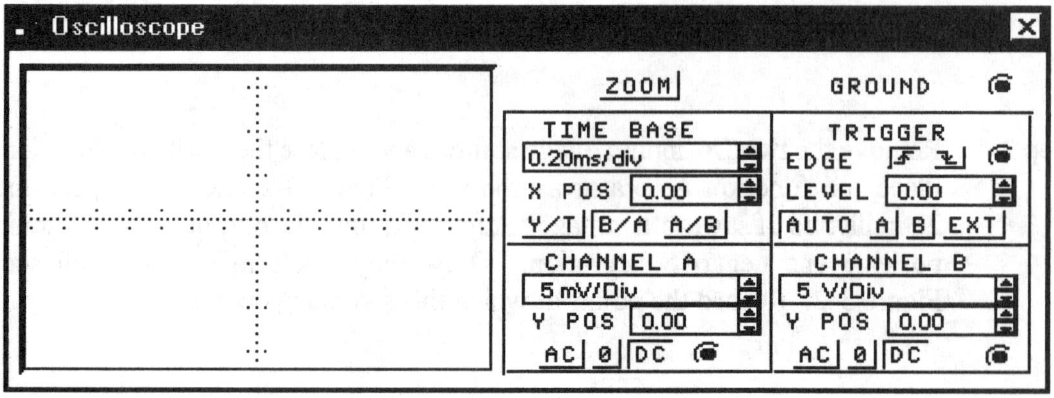

Figure 3-3

Step 4. From the design formulas in the **Key Concepts** section of this chapter design a voltage level detector with the following parameters:

V(UT) = 10 V V(LT) = 4 V Vsat = ±15 V

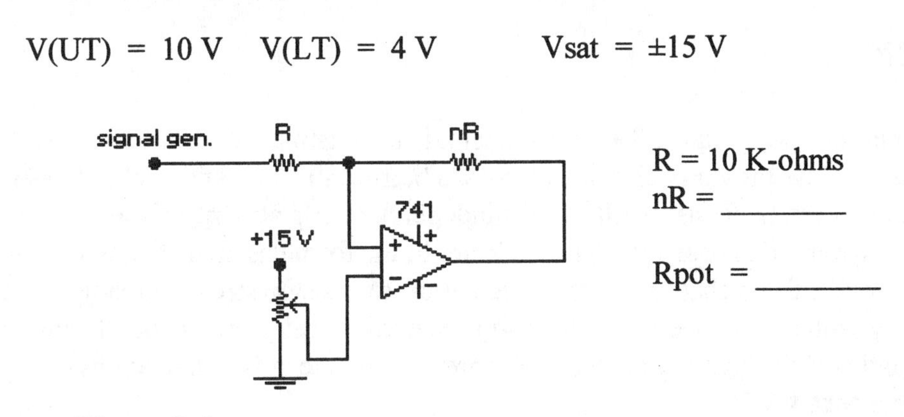

R = 10 K-ohms
nR = ________

Rpot = ________

Figure 3-4

Draw and label the oscilloscope waveform showing Vo versus Vpot on the oscilloscope face below, Figure 3-5.

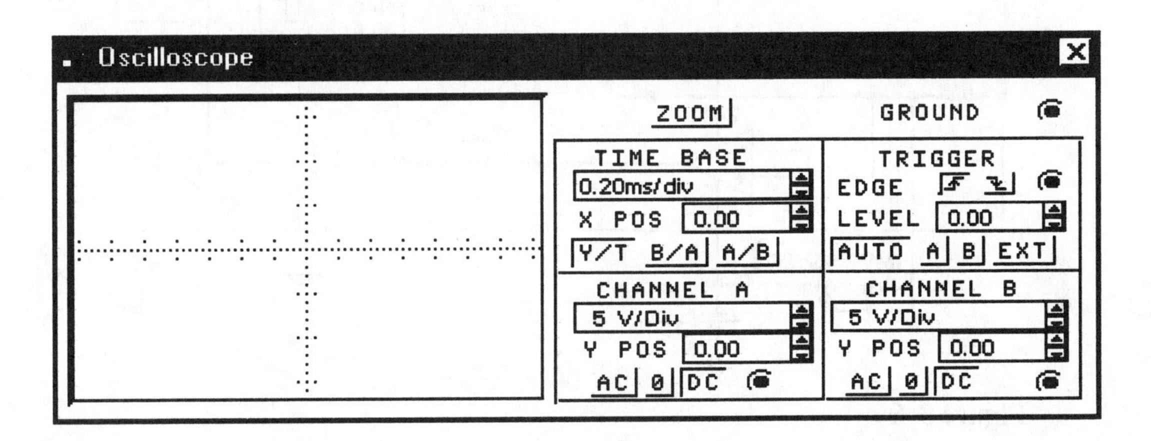

Figure 3-5

Step 5. Replace nR with a rheostat of approximately 100 K-ohms. Explain what happens to the center voltage V(ctr) and the hysteresis voltage V(H).

Exercise 3-2, Independent Adjustment Comparator

PROCEDURE

Step 1. The circuit below, Figure 3-6, has independent adjustment of hysteresis, V(H) and center voltage. The previous circuit did not. The rheostat mR and -V establish the center voltage, V(ctr). Resistor nR allows independent adjustment of the hysteresis voltage V(H), symmetrically around V(ctr). Refer to the formulas in the **Key Concepts** sections at the end of this chapter for the calculation of resistor sizes. To simulate a depleting battery voltage, reduce the battery voltage until the relay drops out, closing the contacts to the theoretical battery charger. Record the voltage at which it begins charging. Load the file "**ex#3-2**."

To begin charge V = ________

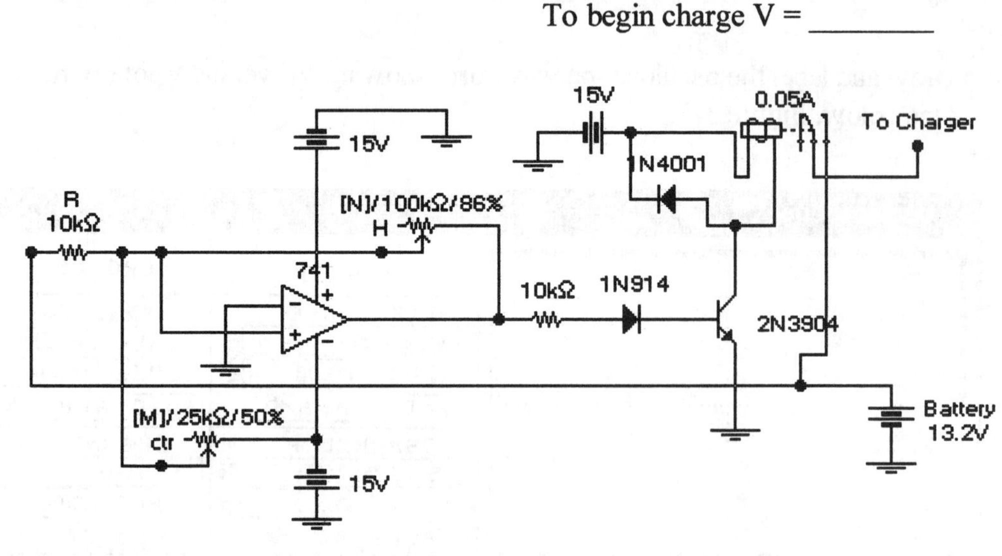

Figure 3-6

Step 2. Adjust the battery upward until the battery has reached full charge (above 13 volts). Record the voltage at which the charging action stops.

Stop charging V = ________

Step 3. From your measurements above, what is the hysteresis voltage of this circuit?

V(H) = ________

Step 4. From your measurements above, what is the center voltage?

V(ctr) = _______

Step 5. From the design formulas, build and simulate a battery charger controller which must charge a 9-volt rechargeable battery. Assume that at 8 volts the battery is not functional and needs a recharge. The charging will continue until the battery reaches 9.5 volts, at which time the relay will disconnect the charger. Let R equal 10 K-ohms. Record your calculated resistor values below. Test your circuit design on *Electronics Workbench.*

V(H) = _______ Vctr = _______

mR = _______ nR = _______

Step 6. **Battery Charger Controller Hardware Circuit Construction**, see Key Concepts. The simulated circuit you used in Step 5 may be used as a starting point.The battery charger controller circuit can be constructed in the laboratory only if the following precautions are observed.

1. A rechargeable battery is used and instruments are connected to warn the experimenter when the maximum charging voltage is reached, so that if the circuit fails it can be shut off manually.

2. The charger being used is approved by the instructor.

Battery Charger Controller Design- Choose a small rechargeable battery to recharge, for example, a 9-volt rechargeable. Obtain manufacturers data on the no-load voltage for a new battery of this type. Other data such as charging rate should be obtained. Record this information below.

Step 7. Construct the battery charger controller circuit. The transistor you choose must be able to supply the current to the relay that you select. Since you are charging a small rechargeable battery, the charging current is going to be in the milliampere range so the relay contact current need not be more than 1 ampere. Choose a small DC relay of not more than 12 volts to energize. Record the specifications of your relay below.

Step 8. Test your charger controller circuit. Your charger will be a laboratory DC power supply with a current limit control adjustment potentiometer on the instrument. Demonstrate that your circuit works to your instructor. Turn in a completed schematic of your battery charger controller and a printout with a parts list and description using EWB software.

Exercise 3-3, Window Comparator Applications

The window comparator can be used to detect voltages within a narrow range, or voltages higher or lower than the range. The LM 311 comparator is a precision comparator that is much more stable than the 741 op-amp. The first example shown below, Figure 3-7, detects a very small difference between a reference voltage and a signal voltage. It will visually indicate when the signal is equal to the reference voltage (within 100 mv); when the signal is higher than the reference voltage, or when the signal is lower than the reference voltage. Examine the circuit below and read the **Key Concepts** section at the end of this chapter.

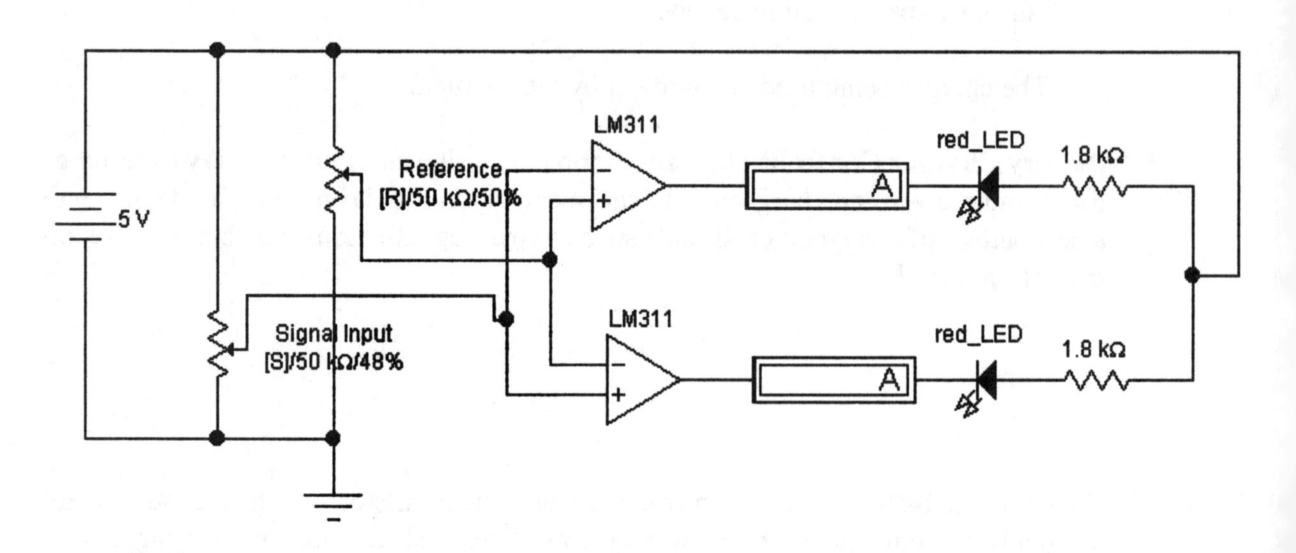

Figure 3-7

Step 1. Load the file labeled "**ex#3-3.**". Note the position of the reference and signal input potentiometers. Place two voltmeters, one at the center of each potentiometer, with respect to ground. Measure and record the voltages.

V(ref) = ________

V(sig) = ________

Step 2. Based on your measurements, what is the voltage difference between the input terminals of each comparator?

V(diff) = ________

Step 3. Which LED is turned on for your settings in Step 1? (upper or lower)

____

Step 4. Set both potentiometers to 50%. Which of the LED'S is turned on? What amount of voltage difference is now between the inputs of each comparator?

LED on? ____________

Voltage (diff) __________

Step 5. Set the signal potentiometer to 52%. Which LED is turned on? What is the voltage difference between the inputs of each comparator?

LED on? ____________

Voltage (diff) __________

Step 6. Modify the circuit as shown in Figure 3-8. Examine the circuit and record the upper and lower voltage limits of the window.

upper limit V = _______

lower limit V = _______

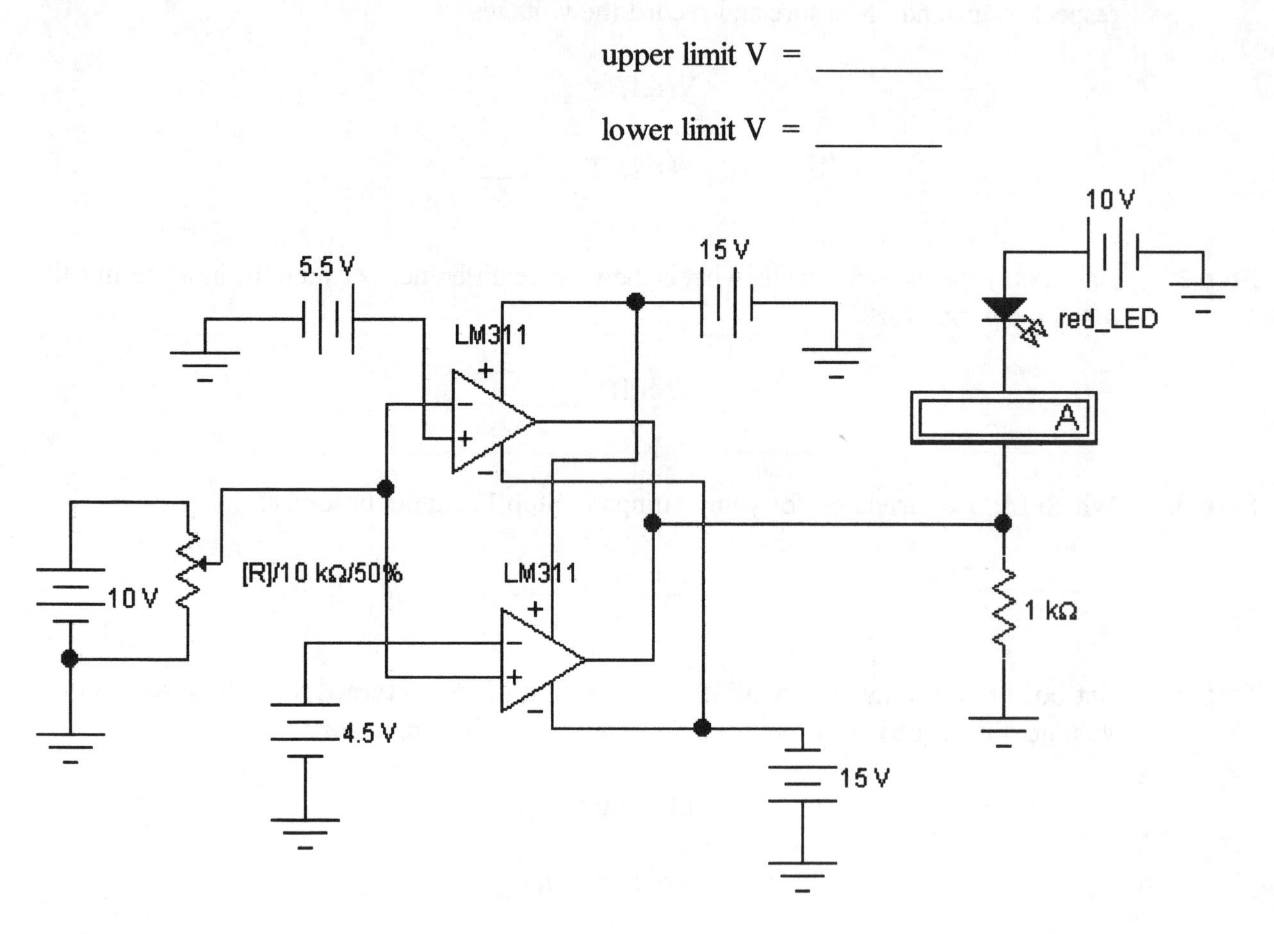

Figure 3-8

Step 7. Move the wiper on the potentiometer by pressing the key R to find the upper trigger point, and (shift) R to find the lower trigger point at which the LED turns on. Place a voltmeter from the wiper of the potentiometer to ground to measure the trigger voltages. Record.

V(UTP) = _______

V(LTP) = _______

Step 8. Place the function generator in series with the voltage coming off the wiper of the potentiometer to the comparator input. Set the function generator to one volt peak and the output to a triangle wave of 100 hertz. Replace the LED with the buzzer. If necessary, adjust the output resistor until the buzzer works. Connect the oscilloscope

CH A to measure the waveform at the comparator input. Connect CH B at the comparator output. Record the upper and lower trigger voltages seen on the oscilloscope. Show the waveforms on the oscilloscope face below, Figure 3-9.

V(UTP) = _______ V(LTP) = _______

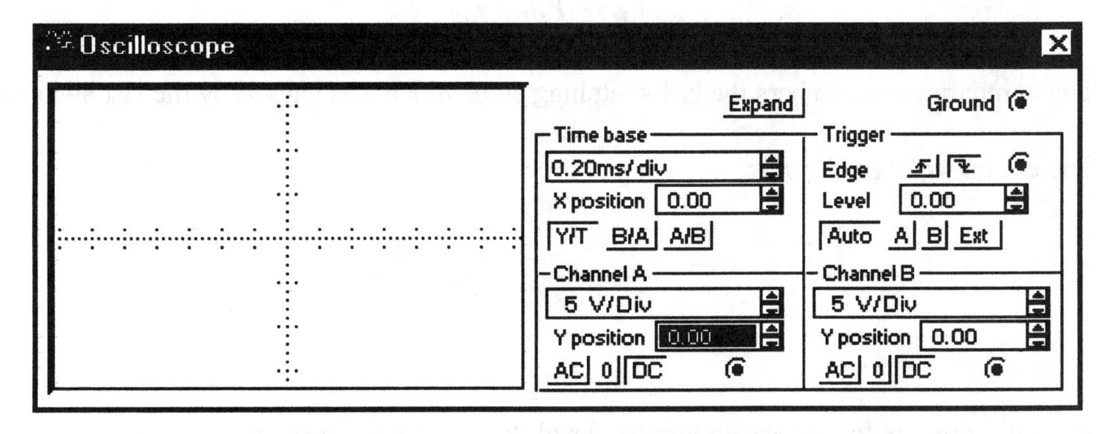

Figure 3-9

Key Concepts
Comparator Circuit Applications

Positive Feedback. Noise can be a major problem in a zero crossing detector if positive feedback is not used. Noise (line or external) can cause false switching and sometimes oscillation.

$$V_{UT} = \frac{R1}{R1 + RF} \times V_{(+\,sat)} \qquad V_{LT} = \frac{R1}{R1 + RF} \times V_{(-\,sat)}$$

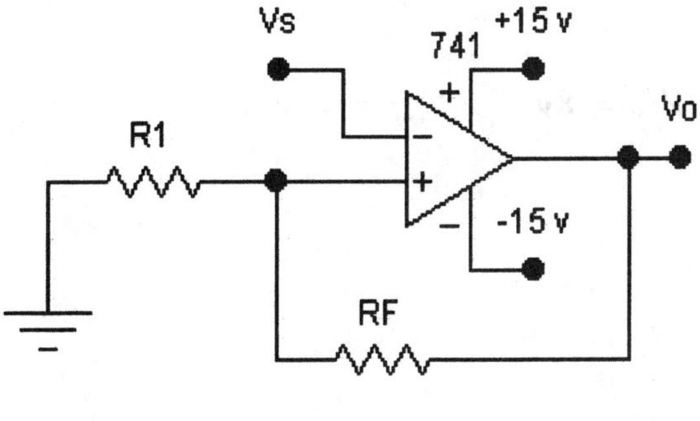

Figure 3-10

Comparator Circuit Applications

If the upper threshold and lower threshold voltages are much larger, then noise voltage false switching is eliminated.

Hysteresis. The difference hysteresis is created from the switching of an output from the first state to a second state and then back to the first state by a changing input signal.

$$V_H = V_{UT} - V_{LT}$$

The comparator remembers the last switching state which was caused by the last switching signal.

The **center voltage** V(ctr)is

$$V_{CTR} = \frac{V_{UT} + V_{LT}}{2}$$

Design formulas for inverting voltage level detector with hysteresis.

Circuit:

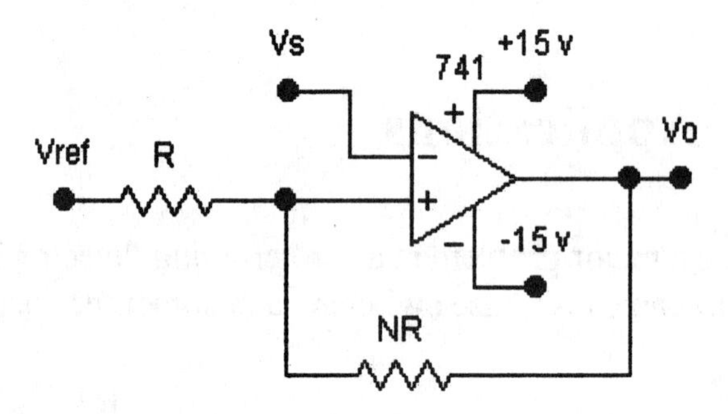

Figure 3-11

Given:

$$V_{UT} = 10v \quad V_{LT} = 8v \quad \pm V_{sat} = 14v$$

Solution:

$$V_{CTR} = \frac{10v + 8v}{2v} = 9\ volts \qquad V_H = 10v - 8v = 2\ volts$$

$$N = \frac{+V_{sat} - (-V_{sat})}{V_H} = \frac{14 - (-14)}{2} = 14$$

Choose R = 10 K then NR = 14 × 10 K = 140 K-ohms

$$V_{ref} = \frac{N+1}{N}(V_{CTR}) = \frac{14+1}{14}x(9) = 9.64\ volts$$

Design formulas for non-inverting voltage level detector with hysteresis.

Circuit:

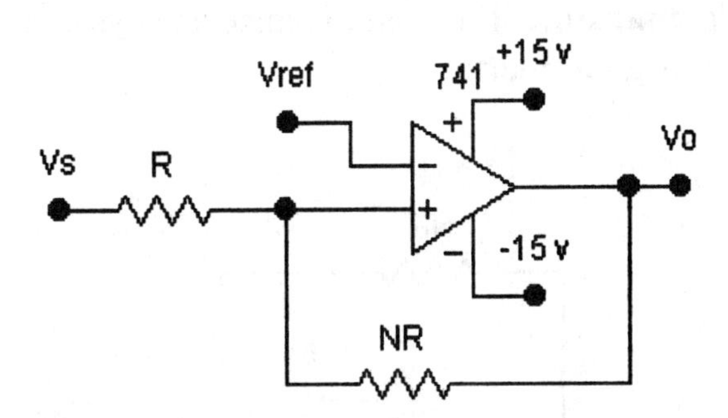

Figure 3-12

Given:

$V_{UT} = 10v \qquad V_{LT} = 8v \qquad \pm V_{SAT} = 14v$

Solution:

$$V_{CTR} = \frac{10v + 8v}{2} = 9\ volts \qquad V_H = 10v - 8v = 2\ volts$$

$$N = \frac{V_{+sat} - (-V_{-sat})}{V_H} - 1 = \frac{14 - (-14)}{2} - 1 = 13$$

$$V_{ref} = \frac{V_{CTR}}{1 + (1 / N)} = \frac{9}{1 + (1 / 14)} = 8.4\ volts$$

Choose R = 10 K then NR = 13 × 10 K = 130 K-ohms

Independent Adjustment Comparator. The center voltage and hysteresis voltage can be adjusted independently of each other. Design formulas.

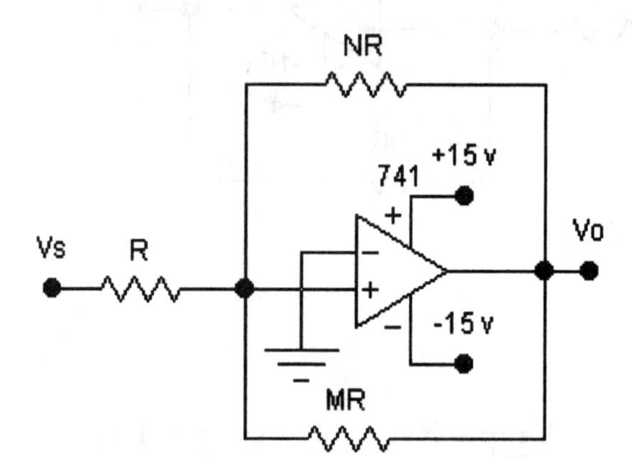

Figure 3-13

Given:

$$V_{UT} = 10v \quad V_{LT} = 8v \quad \pm V_{SAT} = 14\ volts$$

Solution:

$$V_H = 10v - 8v = 2 \text{ volts} \qquad V_{CTR} = \frac{10v + 8v}{2} = 9 \text{ volts}$$

$$M = -\frac{(-V_{ref})}{V_{ctr}} = -\frac{(-15v)}{9v} = 1.66$$

$$N = \frac{V_{sat} - (-V_{sat})}{V_H} = \frac{14 - (-14)}{2} = 14$$

Choose R = 10 K-ohms then MR = 1.66 × 10K = 16.6 K-ohms

and NR = 14 × 10K = 140 K-ohms

Battery charger controller: Hardware design.

The formulas to solve for M(R) and N(R) are shown in the above design. Remember that M(R) is connected to Vref (−V), and adjusts the center voltage while N(R) adjusts the hysteresis (Vut − Vlt). Test this circuit without the transistor circuit and battery charger attached. The battery voltage for this test is the nominal voltage of the battery. This nominal voltage will be the center voltage, for example, 9 volts. The circuit is shown below, Figure 3-14.

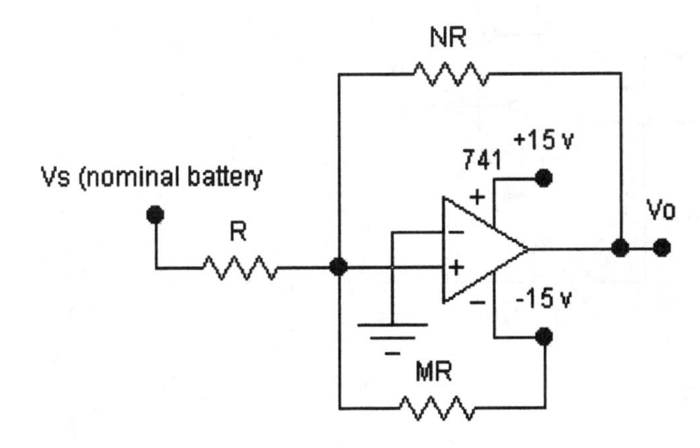

Figure 3-14

Comparator Circuit Applications

Measure Vo vs Vs. Vary Vs to see upper and lower voltage transition points.

Window comparator. Window comparators are often used when one wants to know if input signals are either above or below the window settings. Generally, it is desirable to be within the window limits. In manufacturing it may be necessary to know when a measurement is out of tolerance. The window is created by having the following:

reference voltage - usually set for the measurement of the ideal part dimension

signal voltage - transducer voltage from part measured

The window comparator output visually indicates when the difference is nearly zero (within tolerance). The LED showing the part is good is turned on. When the tolerance is exceeded either in a positive or negative direction the "bad part LED" is turned on.

Chapter 3 Questions
Comparator Circuit Applications

1. Calculate V(ut) and V(lt) for the circuit below. V(ut) = ________

V(lt) = ________

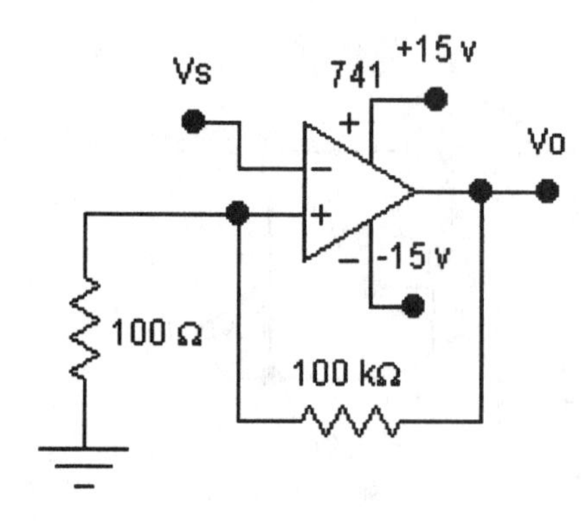

Figure 3-15

2. The circuit in Figure 3-15 is called a(n) ________________________________.

3. _______________ feedback is used in a comparator circuit for the purpose of

____

4. Hysteresis voltage is the difference between ____________ and ___________

5. Center voltage for the battery charging control circuit of a car battery would be

___________ volts.

6. The circuit below is called a(n) ____

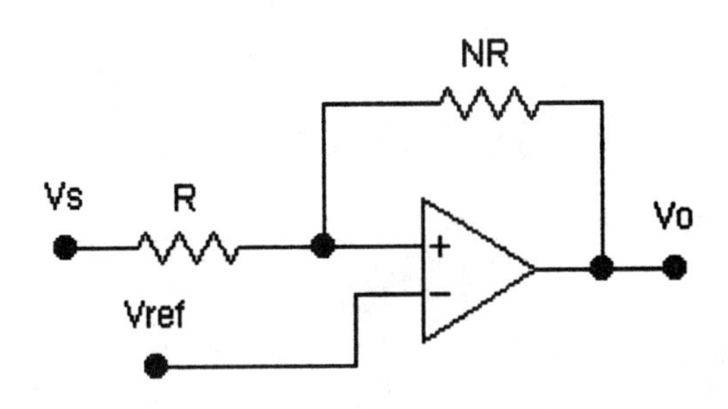

Figure 3-16

7. The LM311 makes a better comparator because it _________________________

Use Figure 3-17 for questions 8 and 9.

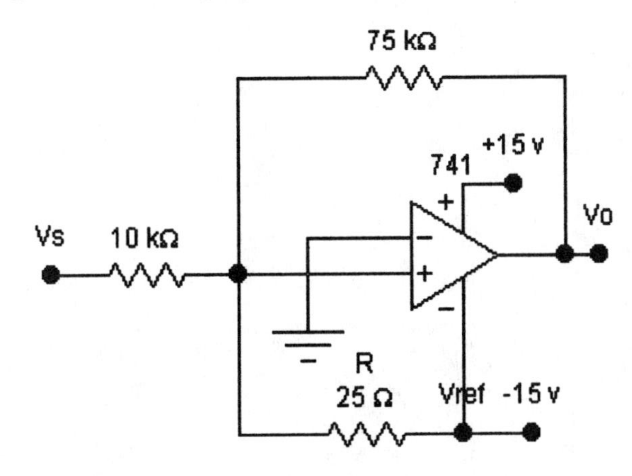

Figure 3-17

8. The center voltage is __________ volts.

Comparator Circuit Applications

9. Upper trigger voltage is ________ volts

   Lower trigger voltage is ________ volts

10. In the window detector lab exercise the light (or buzzer) goes on when the input voltage is (circle one) inside / outside the window area.

CHAPTER 4
ACTIVE FILTER APPLICATIONS

Name ______________________

INTRODUCTION

An active filter using an operational amplifier is commonly used to replace those filters that previously used LC networks. There is a reduction in size and the added benefit of an adjustable voltage when using the active filter.

Low gain circuits provide excellent stability, and, if operated in the non-inverting mode, the RC components are isolated from the load due to the high input impedance of that configuration. This ensures that the RC components will generally determine the corner frequency (also called break frequency and roll-off frequency), while changing the RC ratio or varying the gain will change the waveshape.

A filter is a circuit that will pass a specified band of frequencies while reducing the voltage to the load for all frequencies outside the band. The four types of active filters you will examine in this exercise are: low-pass, high-pass, band-pass, and band-reject (notch filter) filters. You will then examine some applications of active filters in the audio and radio frequency spectrums.

Reading the **Key Concepts** section of this chapter prior to beginning the simulation exercises will help you in understanding the filter design principles and formulas used in the completion of the exercises.

OBJECTIVES

- Gain an understanding of how R and C affect corner frequency.
- Use the EWB bode plotter to produce the frequency response curves of low-pass, high-pass, band-pass and band-reject filters.
- Use the design formulas to calculate the corner frequency of a single pole filter.
- Build and test a filter design to determine if it meets specifications.
- Use the EWB bode plotter to measure phase shift in an active filter circuit.
- From your measurements, calculate Q, bandwidth, and resonant frequency of a band-pass filter.
- Design, build, and test a tone control circuit for specified low and high frequencies.

Exercise 4-1, The Low-Pass Filter

PROCEDURE

Step 1. Load the file labeled "**ex-#4-1a**" from your EWB data disk. Set the BODE PLOTTER controls as instructed in the **Key Concepts** section of this chapter. Connect the bode plotter input to the generator and connect the bode plotter output to the load resistor. Turn the circuit on and then draw and label the frequency response plot on the bode plotter screen below, Figure 4-1. You may move the Y-axis line to find the approximate 3 dB roll-off at the corner frequency by either clicking the left mouse button at the right arrow key on the bode plotter or by clicking and dragging the Y-axis line to the corner frequency while observing the dB change and frequency change on the plotter.

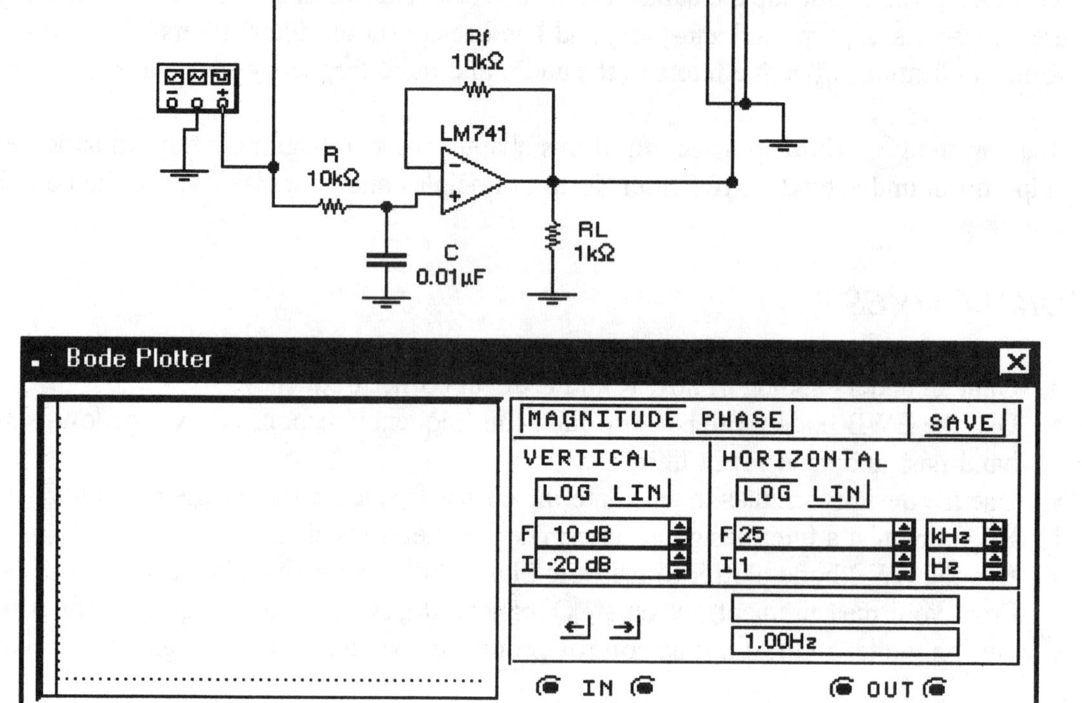

Figure 4-1

Step 2. Bring an oscilloscope to the work space and measure the voltage gain of the circuit. Record the voltage gain. Note: The function generator must be connected for the bode plotter to work on EWB.

Av = ________

Step 3. From the oscilloscope waveshapes of input and output voltage, what is the phase shift of this filter circuit?

Circuit phase shift = ________
@ 3dB corner = ________

Step 4. Switch the bode plotter so that it does a phase plot. From the phase plot shown, what is the circuit phase shift? Note the settings on the bode plotter and the connections from the bode plotter to the circuit. You may need these for future reference when doing other exercises.

Circuit phase shift = ________

Step 5. Click the mouse to open the **Windows** menu at the top left of your screen. Click the word **Description** to open the text window to your screen. From the formula given, design a low-pass filter that has a corner frequency of 1 kHz. Choose a capacitor size of 0.02 uf, or one close to this, that is in your hardware parts supply (your real parts supply). Simulate this circuit on EWB. Draw your circuit design below. Show the part values on your neatly drawn schematic.

Draw the magnitude plot shown on the bode plotter, Figure 4-2. Label the corner frequency and show its location on the plotter screen below. Show your bode plotter settings.

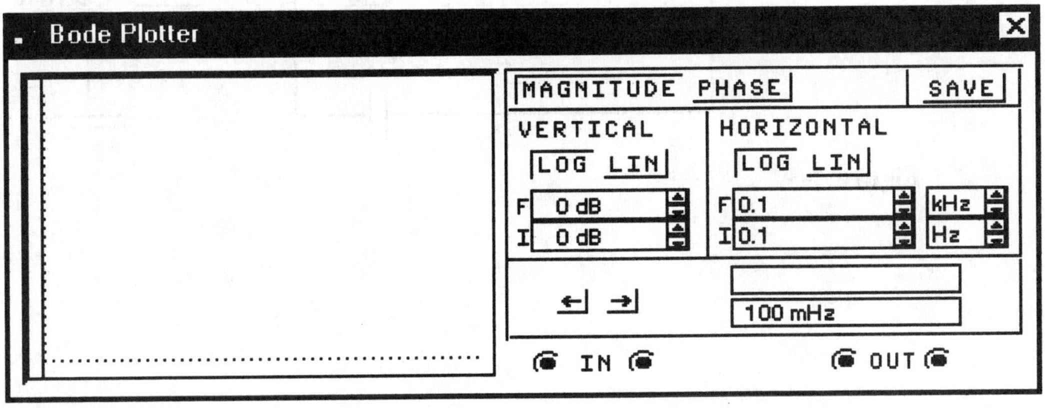

Figure 4-2

Step 6. Modify your design circuit to get a voltage gain of 5. (See your previous non-inverting amplifier circuit exercises if you do not remember how to construct this circuit.) Leave resistor R at its previous value and choose the other resistors so that the DC resistance seen at each op-amp input is approximately equal. Show your calculations below. Also, draw the schematic showing all components with labels and values.

Step 7. Now open the file labeled **"ex#4-1b"**. This is a second-order low-pass filter and has a 40 dB per decade roll-off at the corner frequency. From the **Key Concepts** section of this chapter, determine the size of resistors R1 and R2 that are needed for a corner frequency of approximately 700 Hz.

R1 = R2 = _______

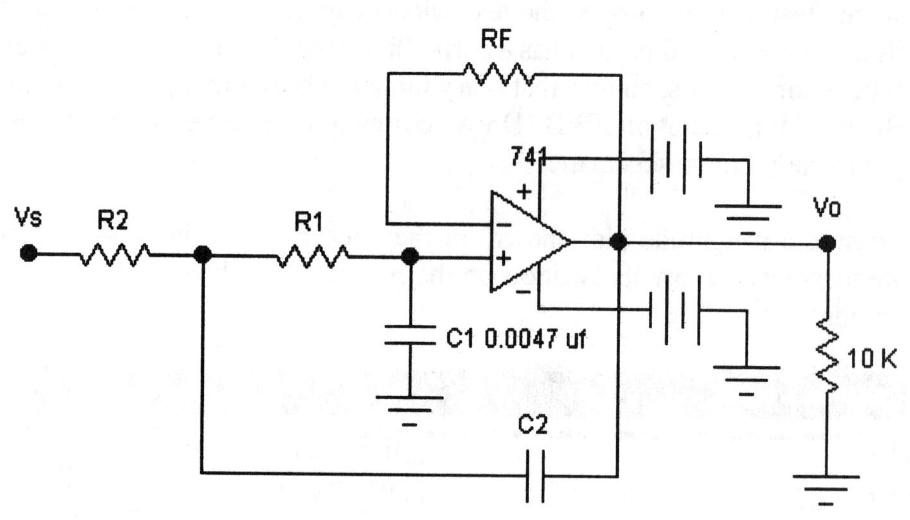

Figure 4-3

Step 8. Place the calculation value for R1 and R2 on the circuit schematic. Connect the bode plotter and signal generator to the circuit. Set the bode plotter controls and then measure the corner frequency. Draw the frequency response on the plotter face below, Figure 4-4. Label the frequency at the −3dB point.

Fc = _______

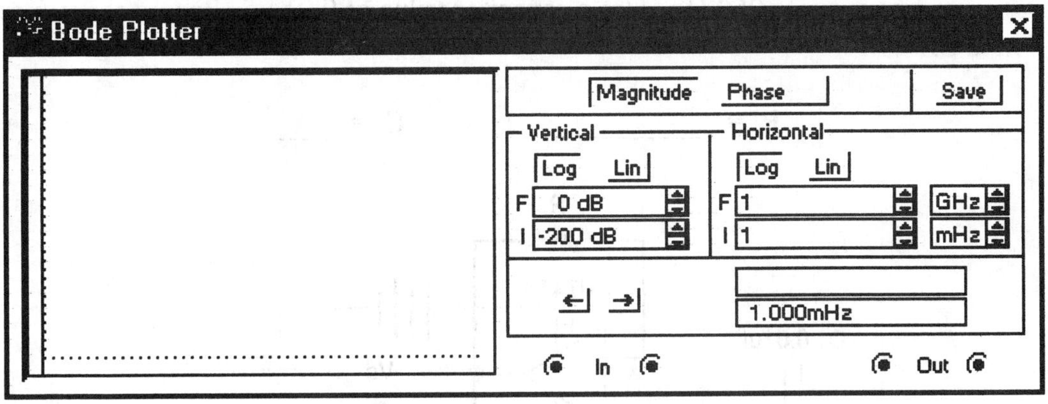

Figure 4-4

Step 9. Modify the previous circuit to have a corner frequency of 1200 Hz. Choose a value of 0.01 for capacitor C1. Solve for C2 and resistors R1, R2, and RF.

R1 = _______ R2 = _______ RF = _______

Step 10. What dB gain is indicated by the bode plotter in the circuit above at approximately one-tenth of the corner frequency? What is the dB gain at the corner frequency? What is the dB gain at twice the corner frequency? What is the dB gain at ten times the corner frequency?

dB @ 0.1 Fc = _______ dB @ Fc = _______

dB @ 2 Fc = _______ dB @ 10 Fc = _______

Exercise 4-2, The High-Pass Filter

PROCEDURE

Step 1. Load the file labeled "**ex#4-2**" from your data disk. Connect the signal generator and bode plotter to the circuit. Calculate the value of the resistors (R) needed to set the high corner frequency to 250 Hz. Use a capacitor value of 0.01 uf. Record the values of R and C.

R = _______ C = _______

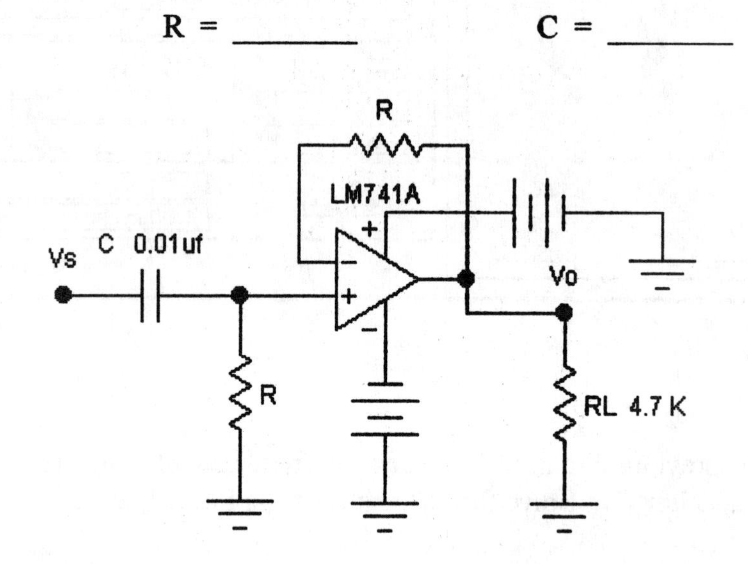

Figure 4-5

Step 2. Measure and record the circuit phase shift from input to the output using the bode plotter.

Phase shift = ______

Step 3. Modify the circuit as shown in Figure 4-6. Measure the voltage gain of this circuit.

Av = _______

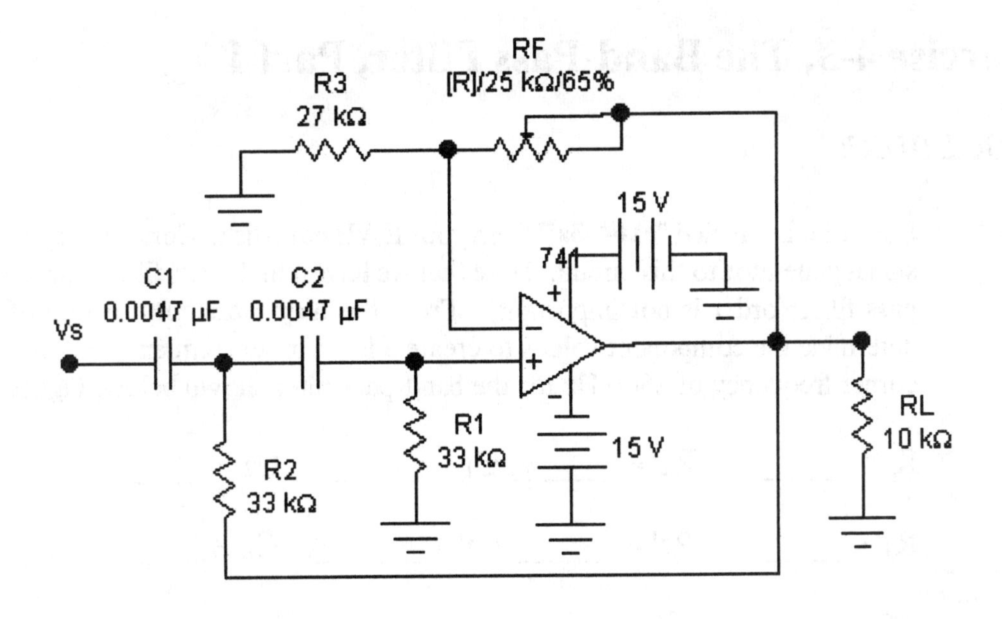

Figure 4-6

Step 4. From your measurement of voltage gain, calculate the voltage gain in decibels.

dB = ________

Step 5. Does this filter circuit meet the approximate specifications for a Butterworth response? Check this out by examining the bode plot response curve. Your instructor may want you to draw this on log-log graph paper. What is rate of roll-off of this filter circuit?

dB(roll-off) = ________

Exercise 4-3, The Band-Pass Filter, Part I

PROCEDURE

Step 1. Load the file labeled "**ex#4-3a**" from your EWB data disk. Connect the bode plotter and signal generator to this circuit. Note that we have a high-pass filter cascaded with a low-pass filter (order is not important). From the **Key Concepts** section of this chapter, determine the component values to create a low corner frequency of 250 Hz and a high corner frequency of 2500 Hz for the band-pass filter shown below, Figure 4-7.

R_1 = ______ R_2 = ______ C_1 = ______ C_2 = ______

R_1' = ______ R_2' = ______ C_1' = ______ C_2' = ______

R_F = ______ R_F' = ______

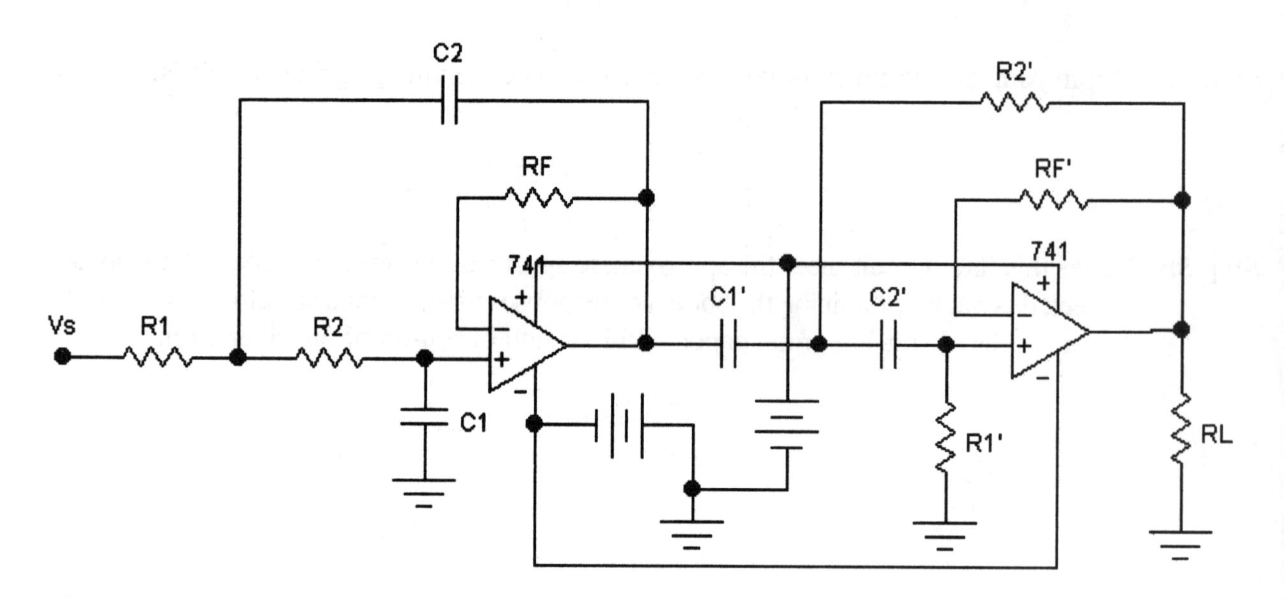

Figure 4-7

Step 2. Place the values in the circuit. If you want to show the values, go to Schematic Options under the Circuit menu.

Step 3. Measure and record the two corner frequency values.

f_L = ________ f_H = ________

Step 4. From your measurements in Step 3, determine center frequency (f r) and the Q of the filter circuit.

f_r = ________ Q = ________

Step 5. Measure and record the following using the oscilloscope and the bode plotter.

What is the voltage gain of the circuit at the center frequency? ________

What is the dB voltage gain of the circuit at the center frequency? ________

What is the dB voltage gain at the low corner frequency? ________

What is the dB voltage gain at one-half the low corner frequency? ________

What is the dB voltage gain at one-tenth the low corner frequency? ________

What is the dB voltage gain at the high corner frequency? ________

What is the dB voltage gain at twice the high corner frequency? ________

What is the dB voltage gain at ten times the high corner frequency? ________

The Band-Pass Filter, Part II, Narrow-Band

PROCEDURE

Step 1. Load the file labeled "**ex#4-3b**" from your data disk. This filter uses a single op-amp but has multiple feedback paths as you can see from the schematic. Also note that the op-amp is inverting as opposed to non-inverting in the previous filters. Design a narrow band-pass filter with a center frequency of 500 Hz with a Q of two.

Make C1 = C2 = 0.1 uf and Rf = 2R. Solve for the unknown resistors.

R = ________ Rr = ________ Rf = ________

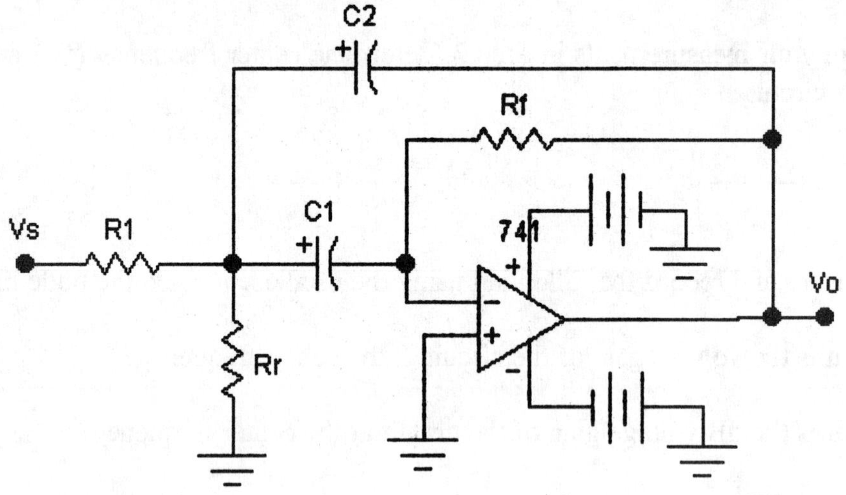

Figure 4-8

Step 2. Measure the resonant frequency, f_r. Record. f_r = ________

Step 3. Measure the low corner frequency, f_L. f_L = ________

Step 4. Measure the high corner frequency, f_H. f_H = ________

Step 5. Measure the voltage gain of the circuit, A_v. A_v = ________

Step 6. What is the circuit bandwidth, BW? BW = ________

Step 7. From your measurement values, what is the circuit Q. Q = ________

Exercise 4-4, The Band-Reject Filter

PROCEDURE

Step 1. The filter circuit shown below, Figure 4-9, is a band-reject circuit which eliminates or notches out frequencies in the reject area. The output of this filter consists of frequencies below and above this canceled-out notch area. Essentially, the notched frequencies are eliminated by the addition of Vs and inverted Vs (−Vs).

Load the file labeled "**ex#4-4**" from your data disk. Design a notch filter for a resonant frequency of 500 Hz and a bandwidth of 50 Hz. Use the same formulas that you used in designing the narrow-band filter in the previous exercise. Let C1 and C2 equal 0.1 uf and RF equal 2R for the band-pass section of the circuit. R2, R3, and R4 are 10 K-ohm, 1% resistors.

R1 = _______ Rf = _______ Rr = _______

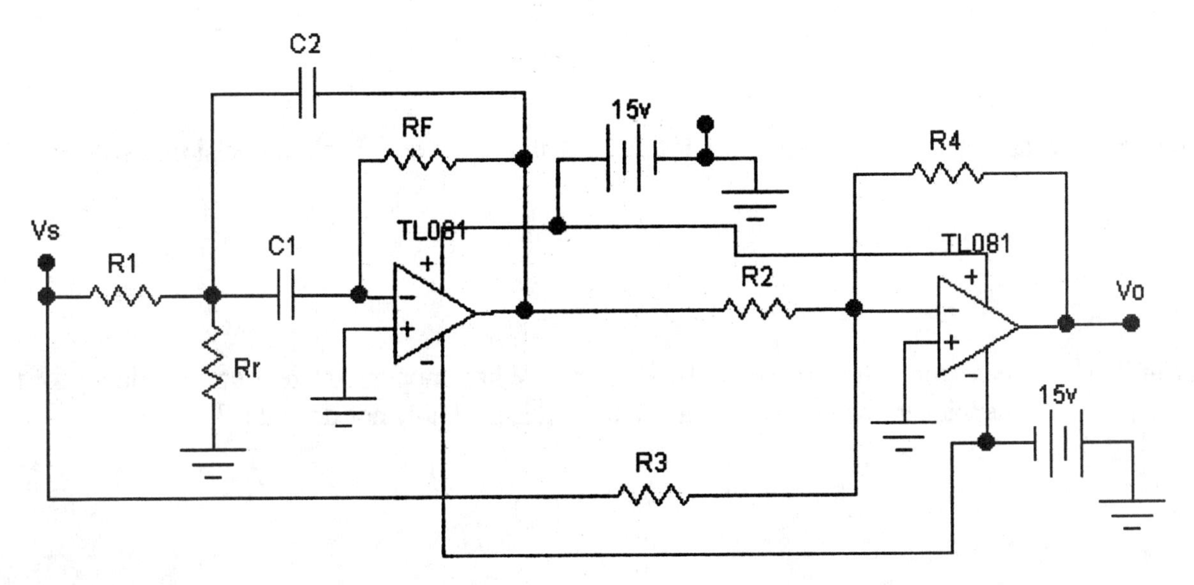

Figure 4-9

Step 2. Connect the signal generator and the bode plotter to the input and output terminals respectively. Set the controls of the bode plotter so that the 0.707 corner frequencies can be measured.

f_L = ________ f_H = ________

Step 3. Measure and record the resonant frequency (point at maximum notching).

f_r = ________

Step 4. Based on your measurements, what is the band of frequencies that this circuit notches out?

BW = ________

Step 5. Make either R2 or R3 adjustable. State the effect on the depth of the notch if the adjustable rheostat is made 10% larger than the original value.

Step 6. State why you think resistors R2 and R3 should be equal value precision resistors.

Step 7. Increase the value of R4 to 15 K-ohms. What happens to the output voltage for the frequencies in the notched area? Why is this probably not desirable?

Key Concepts
Active Filter Applications

Active Filter. An active filter has discrete components such as resistors, capacitors, and occasionally inductors. It also has an active component called an op-amp, which is capable of voltage gain and may be operated in the inverting or non-inverting mode.

Low-Pass Filters. The low-pass filter passes lower frequencies and has a relatively constant output voltage from DC up to a "corner frequency," at which it begins to attenuate or block those higher frequencies from reaching the circuit output terminal.

1st Order low-pass filters attenuate frequencies beyond the corner frequency at a rate of 6dB per or octave (twice the corner frequency), or stated another way, they attenuate at 20 dB per decade (ten times the corner frequency). The input circuit of the filter determines the roll-off rate. The corner frequency is also called "the roll-off," "the 0.707 point," and "the 3 dB point." All these terms are referring to the same principle, as illustrated below, Figure 4-10.

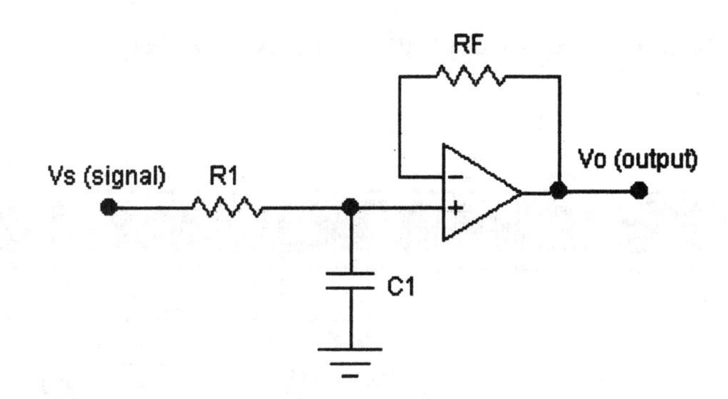

Figure 4-10

When the frequency of the input signal causes X_c to equal the opposition of resistor R1(using the voltage divider rule for AC circuits), the voltage at C1 becomes 0.707 of the input voltage Vs. The voltage at the output of the op-amp is also 0.707 of Vs since it is operating as a voltage follower.

This voltage loss at the corner frequency is:

$$\text{dB (loss)} = 20 \log (0.707 / 1) \text{ or } 20 \log 0.707 = -3 \text{ dB}$$

Therefore,

$$f_c = \frac{1}{2\pi\, R1\, C}$$

and

$$R1 = \frac{1}{2\pi fC} = \frac{1}{\omega C}$$

The frequency response curve of the 1st order low-pass filter is shown below, Figure 4-11. The vertical line shows the corner frequency (0.707 point, −3dB point, roll-off point).

The phase shift from input to output will be 45 degrees at the corner frequency and increase to approximately 90 degrees at 10(fc).

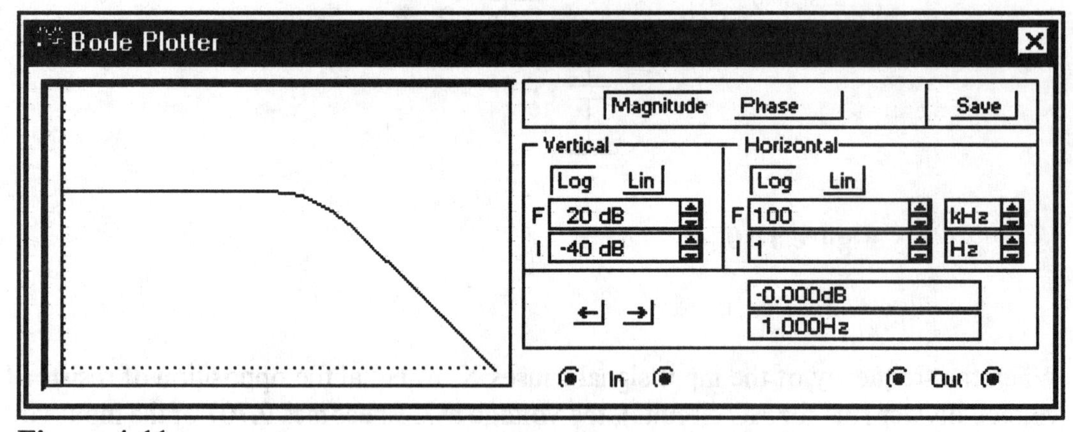

Figure 4-11

First Order Low-Pass Filter Design

Given: You need a **First Order** filter with a corner frequency of 2000 Hz.

1. Choose a common capacitor size in the range of 0.001 uf to 0.01 uf (approximately).

2. Calculate the value of R from the formula:

$$R = \frac{1}{\omega C} = \frac{1}{2\pi f C}$$

3. Solve for R. Choose RF = R.

The **Second Order Low-Pass Filter** has a roll-off of 40 dB per decade or approximately 12 dB per octave. This much steeper roll-off obviously attenuates frequencies above the corner frequency when the feedback capacitor's capacitive reactance becomes smaller and reduces the feedback signal. Note that the op-amp is connected as a non-inverting amplifier; however, the feedback signal does have a phase shift of 90 degrees at the corner frequency (fc), to almost 180 degrees at 10(fc). Therefore, the attenuation becomes steeper as the frequency of the input signal increases. The circuit is shown below, Figure 4-12.

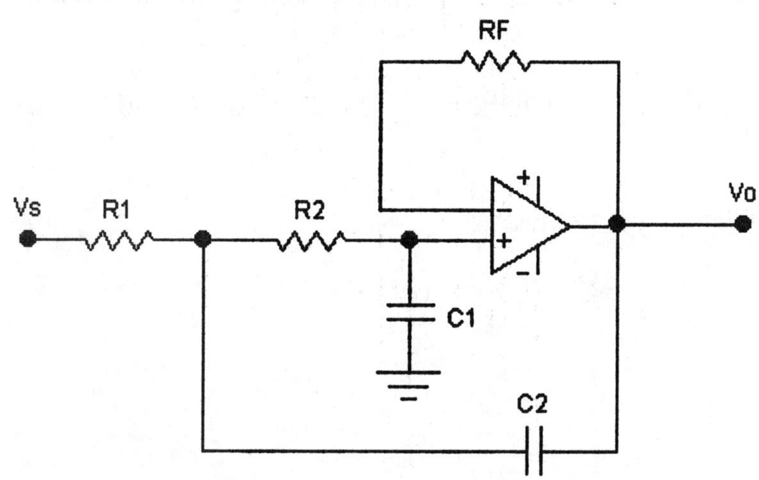

Figure 4-12

The values for R1, R2, C1, C2, and RF are determined as follows:

If, R = R1 = R2, and C = C1 = C2

$$R = \frac{0.707}{2\pi fC1}$$

From the formula above, make R1 = R2 and C2 = 2C1

Then,

$$R = \frac{0.707}{2\pi fC1}$$

$$R_F = 2R$$

Second Order Low-Pass Filter Design

Given: You need a low-pass filter with a corner frequency of 2000 Hz and a roll-off of 40 dB per decade.

1. Choose a common capacitor size (C1) in the range of 0.01 uf to 0.001 uf (approximately).

2. Calculate the value of R from the formula:

$$R = \frac{0.707}{2\pi fC1}$$

3. Solve for the value of R1 and R2, where R = R1 = R2.

4. Solve for the value of C2, where C2 = 2C1.

5. Make RF = 2R.

Note: You may use the alternate design method above.

Uses of low-pass filters:

1. Reduction of ripple in power supplies.
2. Elimination of 60 Hz and 120 Hz hum in audio circuits.
3. Passing low frequencies to audio amplifiers.
4. Passing low frequencies to a speaker.

High-Pass Filters. The high-pass filter passes high frequencies above a designated corner frequency. At frequencies below the corner frequency, the output voltage falls quickly. The input circuit determines the corner frequency.

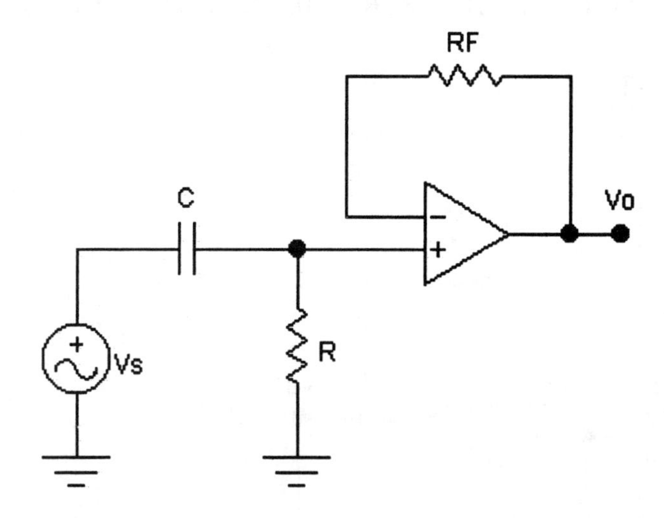

Figure 4-13

For a −3dB roll-off at the corner frequency (6 dB per octave or 20 dB per decade), the formula is:

$$R = \frac{1}{2\pi f_c C} = \frac{1}{\omega_c C}$$

Note: One octave equals 2 × fc, while one decade equals 10 × fc .

Design Example: It is determined that a high-pass filter with a roll-off of 20 dB per decade is needed.

Step 1. Arbitrarily choose a common capacitor size, such as, C = 0.001 uf.

Step 2. Solve for R.

$$R = \frac{1}{2\pi \times 500 \times 0.01 \times 10^{-6}} = 32K - \text{ohms}$$

The *2nd Order High-Pass Filter* has a roll-off of 40 dB per decade from the corner frequency.

If we make Ca = Cb = C, and Ra = Rb = R, and Rf = R (**Method 1**) then our formula remains the same as in the design example above.

An *alternative* method (**Method 2**) is to make Ra = R, Rb = 0.5Ra, and Ca = Cb = C. In this case the formula becomes:

$$R = \frac{1.414}{\omega_c C}$$

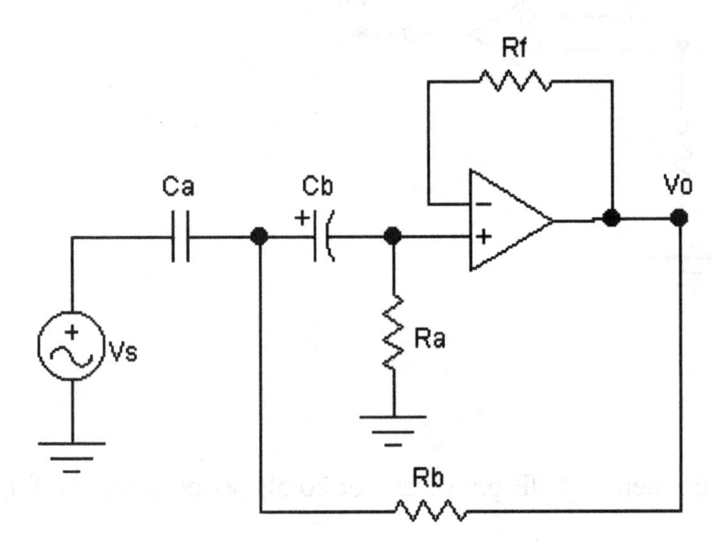

Figure 4-14

Design Example: Using Method 1 above, design a 60 dB/decade high-pass filter with a corner frequency of 500 Hz.

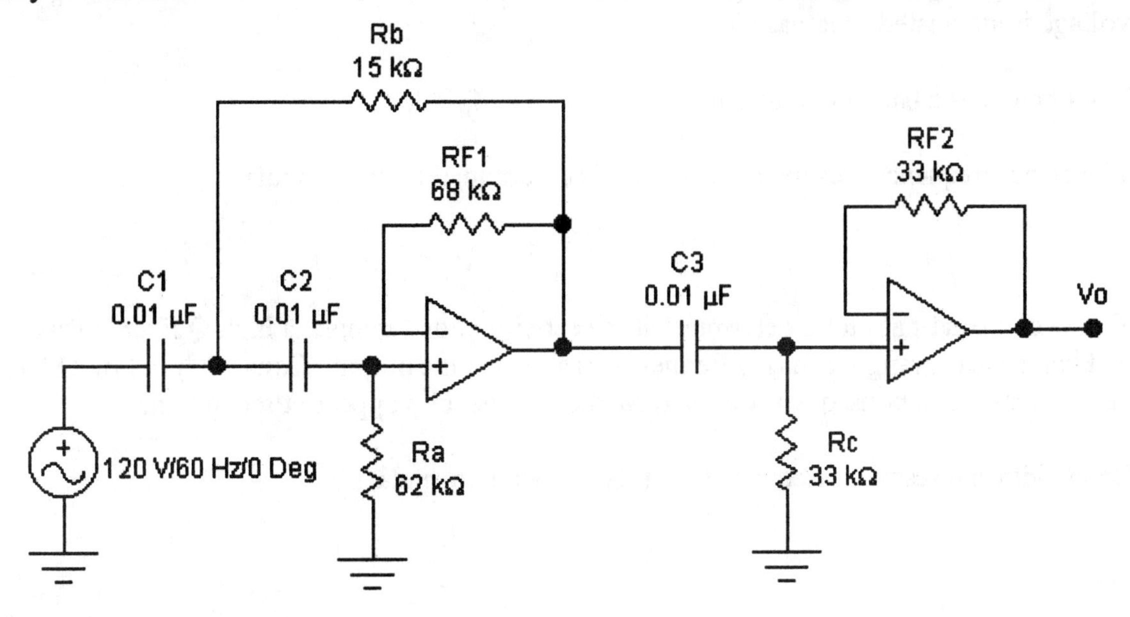

Figure 4-15

Given: C = C1 = C2 = C3 = 0.01 uf, and Ra = 2Rc, and Rb = 0.5Rc. Select the feedback resistors to equalize input bias currents.

Starting with the −20 dB/decade filter,

1. Solve for Rc, where

$$Rc = \frac{1}{2\pi fC} = \frac{1}{2\pi(500)0.01 \times 10^{-6}} = 32\ \text{K - ohms}$$

2. Solve for Ra.

$$Ra = 2Rc = 64\ \text{K - ohms}$$

3. Solve for Rb.

$$R_b = \frac{1}{2}R_c = 16\ \text{K - ohms (use a 20 K - ohm potentiometer)}$$

Band-Pass Filters- The band-pass filter passes frequencies within a range (band) from a low (f_L) corner frequency to a higher frequency (f_H) corner. For frequencies below f_L and above f_H the output voltage is attenuated (decreases)

The circuit has a bandwidth which is $BW = f_H - f_L$

The center frequency (resonant frequency) is determined by the formula:

$$f_r = \sqrt{f_H f_L}$$

The **Q** of a band-pass filter determines its selectivity. For example, a high-Q filter is very selective and has a narrow range (band) of frequencies as seen at the output. Conversely, a low-Q filter is not very selective and consequently a wide band of frequencies appear at the output.

Bandwidth and resonant frequency are related by the formula:

$$BW = \frac{f_r}{Q}$$

A **cascaded** band-pass filter has a low corner frequency determined by the high-pass filter and a high corner frequency determined by the low-pass filter. The maximum voltage gain occurs at the center frequency. The circuit is shown in Figure 4-16.

Design Example (Wide-Band Filter): Choose a corner frequency of 2800 Hz for the low-pass filter. Choose a corner frequency of 500 Hz for the high-pass cascaded section.

1. Solve for the value of R3 and R4 in the high-pass second order filter. Arbitrarily choose C3'=C4'=C and R3'=R4'=R. For example, choose C = 0.02 uf, then solve for R.

$$R = \frac{1}{2\pi fC} = \frac{1}{6.28(500)(0.02\ uf)} = 15.9\ K\text{-}ohms$$

2. Solve for the value of R1 and R2 in the low-pass second order filter. Arbitrarily choose C1=C2=C and R1=R2=R. For example, choose C = 0.01 uf, then solve for R.

$$R = \frac{1}{2\pi fC} = \frac{1}{6.28(2800)(0.01\ uf)} = 5.7\ K\text{-}ohms$$

3. Use either closest common size values or precision potentiometers for more accuracy.

4. Solve for f_r, where:

$$f = \sqrt{f_L f_H} = \sqrt{(500)(2800)} = 1183\ Hz$$

5. Solve for circuit Q, where:

$$Q = \frac{f_r}{BW} = \frac{1200\ Hz}{2800\ Hz - 500\ Hz} = 0.52$$

Design Example (Narrow-Band Filter): This filter is usually constructed with one op-amp. It has two feedback paths and is constructed as an inverted output as shown below.

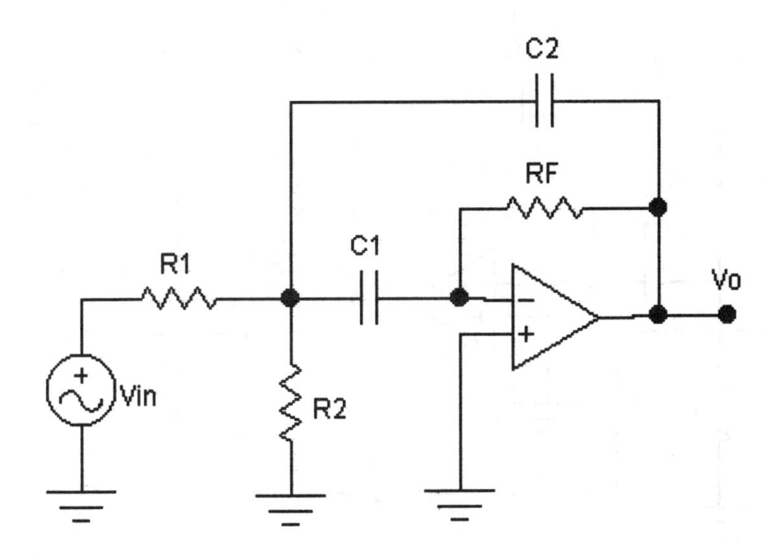

Figure 4-16

Design a 250-hertz unity gain narrow band-pass filter with a bandwidth of 50 hertz.

1.

$$Since\ BW = \frac{f}{Q} \quad and \quad BW = \frac{1}{2\pi R_1 C_1}$$

Then,

$$R_1 = \frac{1}{2\pi(50)(0.001\ uf)} = 64\ K\text{-}ohms$$

2. Make RF equal to 2(R1). R1 = 128 K-ohms.

3. Solve for R2 using the equation:

$$R_2 = \frac{R_1}{2Q^2 - 1} = \frac{64\ K\text{-}ohms}{[2(5^2)] - 1} \approx 1280\ ohms$$

Band-Reject (Notch Filter): A band-reject filter is usually selected to reject unwanted frequencies. These unwanted frequencies are attenuated so they do not interfere with desired operation of electronic equipment. Generally, this is a narrow band of unwanted frequencies. A notch filter can be made by subtracting the output of a band-pass filter from the original signal.

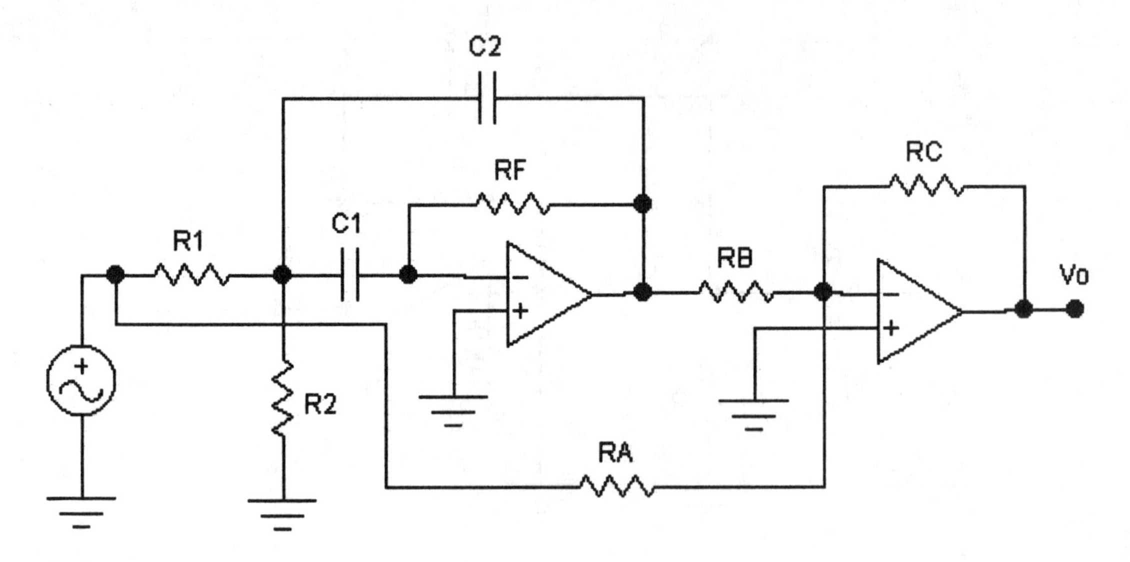

Figure 4-17

Design a 120 Hz notch filter with a bandwidth of 10 Hz. Arbitrarily choose C1 = C2 = 0.33 uf. Make the adder resistors RA = RB = RC = 10 K-ohms. For extreme accuracy use 1% resistors.

1. Solve for circuit Q from the equation:

$$Q = \frac{f_r}{BW} = \frac{120}{10} = 12$$

2. Solve for resistor R1 from the equation:

$$R_1 = \frac{0.159}{(BW)C} = \frac{0.159}{(10)(0.33\,uf)} \approx 48\,K-\text{ohms}$$

3. Solve for R2 from the equation:

$$R_2 = \frac{R_1}{2Q^2 - 1} = \frac{48K}{\left[2(12^2)\right] - 1} \approx 167\ \text{ohms}$$

4. Make R3 equal to twice the value of R1. So R3 = 96 K-ohms.

Chapter 4 Questions
Active Filter Applications

True/False: Put answers on the lines provided below.

1. ______ A low-pass filter only passes frequencies higher than the roll-off frequency.

2. ______ Increasing the capacitor size for a low-pass filter decreases the corner frequency.

3. ______ The roll-off of a second order low-pass filter is steeper than a first order filter.

4. ______ A notch filter attenuates a narrow band of frequencies.

5. ______ A low-Q bandpass filter has a narrow bandwidth.

Active Filter Applications

Fill in the answers to questions 6, 7, and 8 using Figure 4-18.

6. The filter shown in Figure 4-18 is a ______________________________ filter.

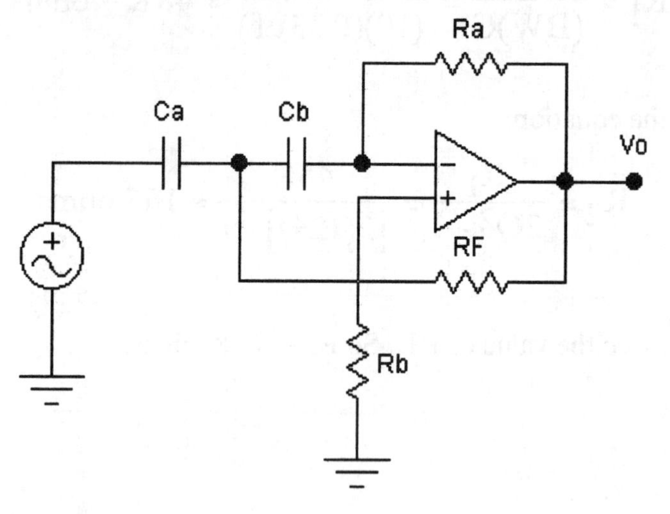

Figure 4-18

7. Decreasing the values of Ca and Cb will ______________________________ the roll-off frequency.

8. What is the corner frequency if Ra=Rb=27 K-ohms and Ca=Cb=0.002 uf?

____

9. A narrow-band filter has a ______________________________ Q.

10. Describe and sketch a method of measuring the corner frequency using laboratory equipment.

CHAPTER 5
SIGNAL GENERATING CIRCUITS

Name ______________

INTRODUCTION

There are special signal generating ICs that will create multiple waveforms off a single chip. The four types of common waveforms are sine, square, sawtooth, and triangular waves. Signal generators are generally known as oscillators. An oscillator is an amplifier with enough positive feedback returned to the input to sustain a continuous waveform at the output. The components in the feedback circuit determine the shape and frequency of the generated wave. The stability of an oscillator (ability to output the desired frequency at a fairly constant amplitude) depends on careful design, temperature control, and power supply stability. In this chapter you will examine various oscillator designs.

OBJECTIVES

- Predict and measure the frequency of common oscillator circuits.
- Predict and measure the output waveshapes of oscillators.
- Measure the voltage gain and phase shift of oscillator circuits.
- Adjust the amount of feedback to control gain and regenerative feedback.
- Learn the start-up conditions for oscillation.

Exercise 5-1, The Phase Shift Oscillator

PROCEDURE

Step 1. Load the file labeled **"ex#5-1"** from your EWB circuits disk. This circuit is the least stable of the oscillators that you will examine. However, it provides a good explanation of oscillator principles. Using the design principles in the **Key Concepts** section of this chapter, design a phase shift oscillator with an output frequency of 400 Hz. Using C = C1 = C2 = C3 = 0.01 uf, what value of resistor, R, is needed? RF?

R = _______

RF = _______

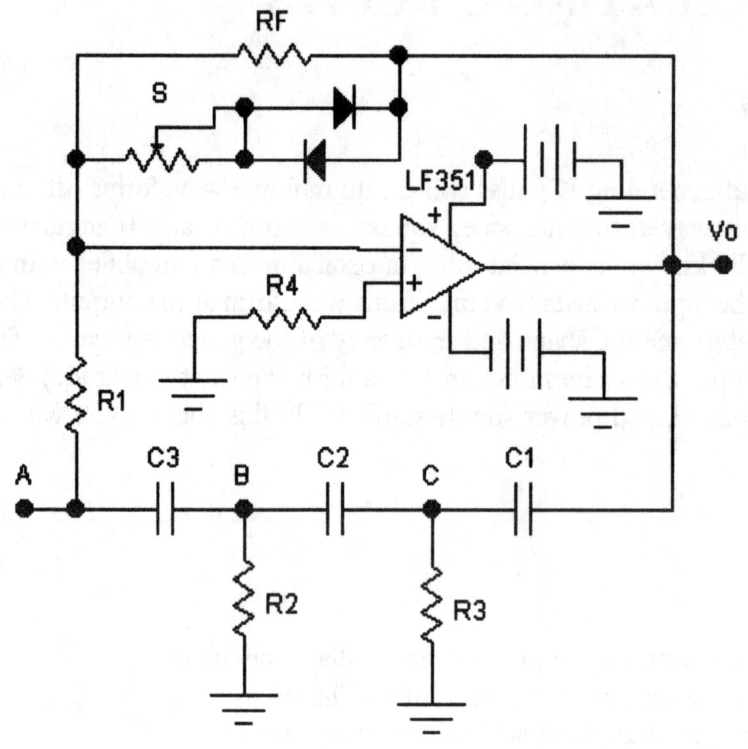

Figure 5-1

Step 2. Place the calculated values in this circuit and then check and record its oscillating frequency. You may need to tweak (adjust slightly) the potentiometer value to start circuit oscillation.

f = _______ Hz

Note: This oscillator requires a kick-start. The diodes and adjustable resistor (S) provide the kick. The parallel combination with RF should be approximately 29 times larger than R1.

Step 3. Measure and record the phase shift of the signal at points A, B, and C with respect to the op-amp output terminal Vo.

Φ_A =

Φ_B =

Φ_C =

Step 4. Connect LED's with current limiting resistors to show each half cycle of the sine wave of oscillation. Limit the current to each diode to about 5 mA.

Step 5. Connect a buzzer or speaker that either does not require much current or has a current driver to supply current to the speaker. You can use the EWB buzzer and set the current to a low value. Print out your working circuit and hand it in with this lab assignment. Include the oscilloscope printout of the oscillator output waveform.

Step 6. Calculate the amount of current required to drive a 1/4 watt, 8-ohm speaker and record below. Consult a data book, vendor catalog, or your local electronics store to find an audible device that operates at a current lower than this speaker. Record the part number, manufacturer, and operating specifications below.

I = _______

Specifications:

Exercise 5-2, The Wien Bridge Oscillator

PROCEDURE

Step 1. Load the file labeled **"ex#5-2"** from your EWB data disk. Based on the **Key Concepts** theory and formulas in this manual, design a Wien oscillator with an output frequency of 1 kHz. Assume C = C1 = C2, R = R1 = R2 = R3, and RF = 2R. Choose a capacitor of C = .01 uf and then solve for R.

R = _______

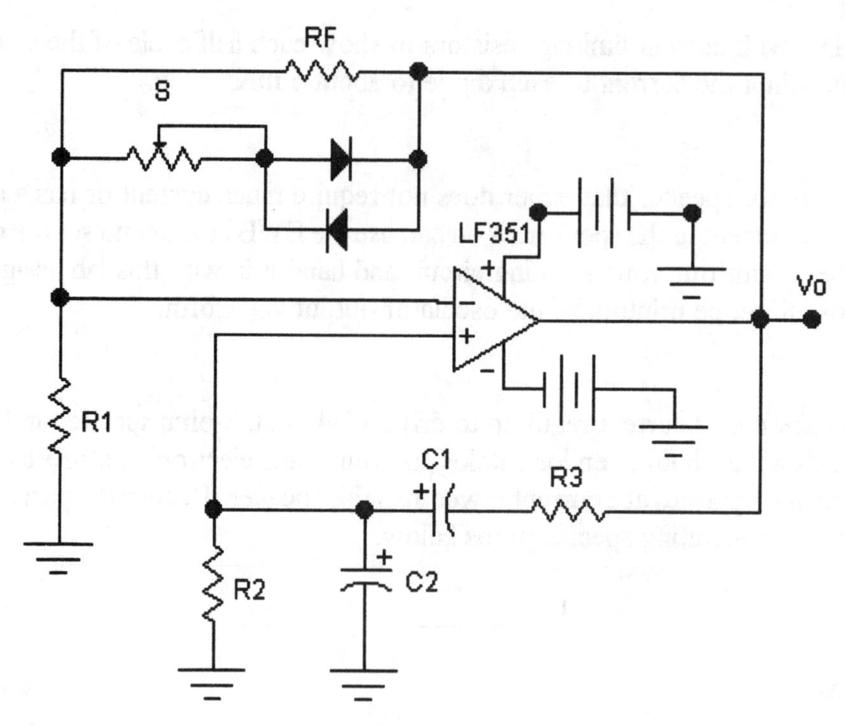

Figure 5-2

Step 2. Add two back-to-back 1N4739 (9.1 volt) zeners at the output terminal Vo. Measure and record the new positive and negative swings of the output voltage. Show the output waveform on the oscilloscope screen below.

Vo p-p = ________

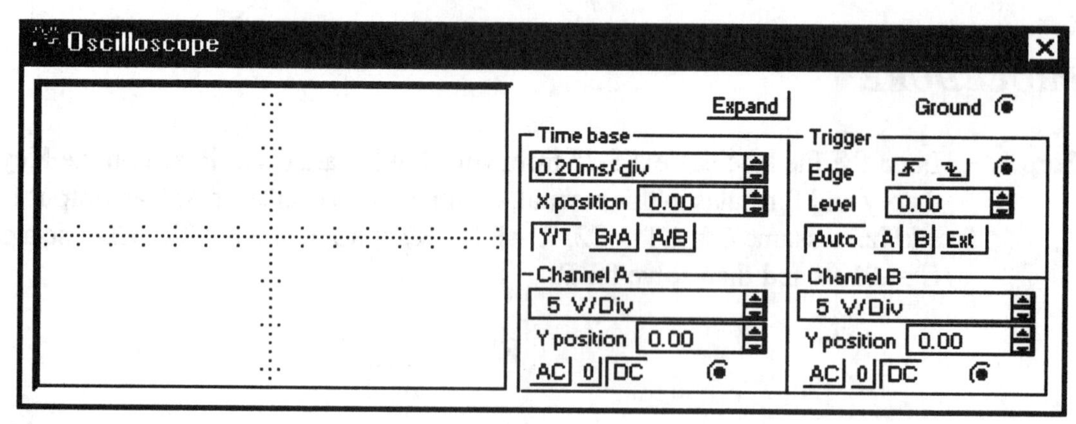

Figure 5-3

Step 3. Attach a transistor driver circuit that will operate a bell/buzzer that requires 9 volts at a current of 50 milliamperes. Connect the EWB buzzer (don't forget to set the buzzer parameters) and test for audible response. Draw the oscilloscope output waveform on the oscilloscope face below. Draw your driver circuit below.

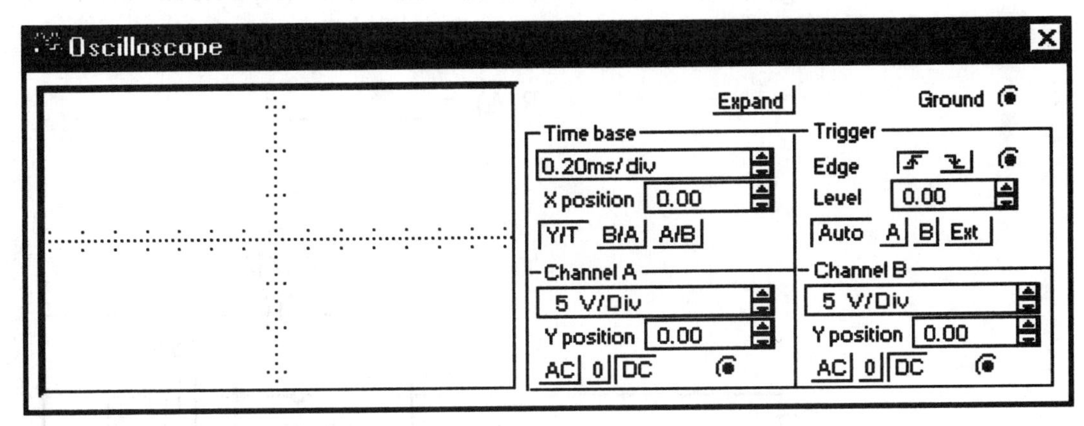

Figure 5-4

Step 4. See if you can design a Wien oscillator to operate at 1 MHz. What are the problems you encountered? Did it work? Draw your circuit schematic with parts identified below. What is the highest frequency (sine wave) you can get?

Exercise 5-3, Triangle and Sawtooth Oscillators

PROCEDURE

Step 1. Load the file labeled "**ex#5-3**" from your EWB circuit files disk. The circuit shown on your screen and below, Figure 5-5, is analyzed in the **Key Concepts** section of this chapter. Read this carefully so that you will understand its operation and design. Your first step will be to set the peak output voltage of the ramp by adjusting the (A) potentiometer of the input comparator stage. Leave the slope potentiometer (S) of the integrator stage set to 50%. Calculate the value of the (A) potentiometer to give a 5-volt peak at the integrator output.

R(A) = _______

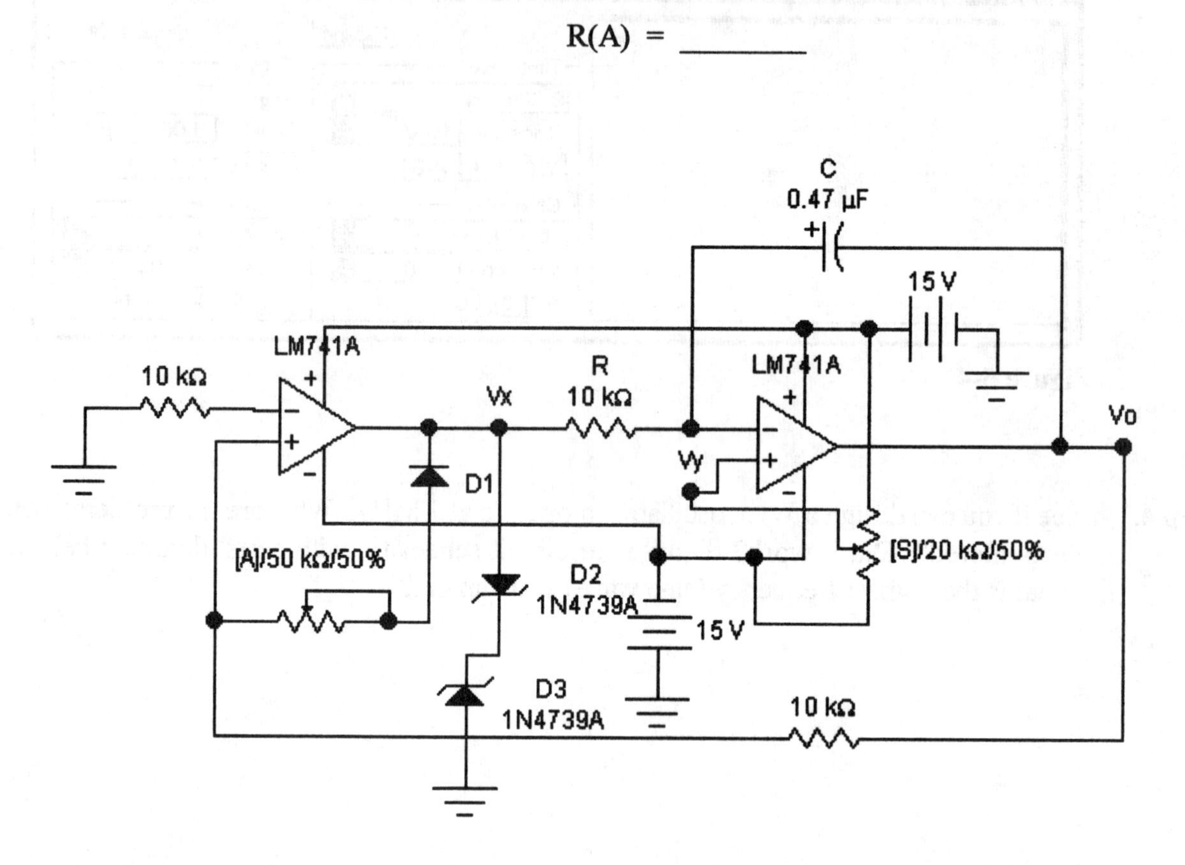

Figure 5-5

Note: To start the oscillator, you may have to adjust (S). Press the S key, then press Shift S to return to 50%. For fine adjustment of (A), double-click at key A.

Step 2. Measure the voltages Vx and Vy and record.

Vx = _______

Vy = _______

Step 3. Draw and label the oscilloscope waveforms you measure at points Vx and Vy. Show these on the oscilloscope face below, Figure 5-6.

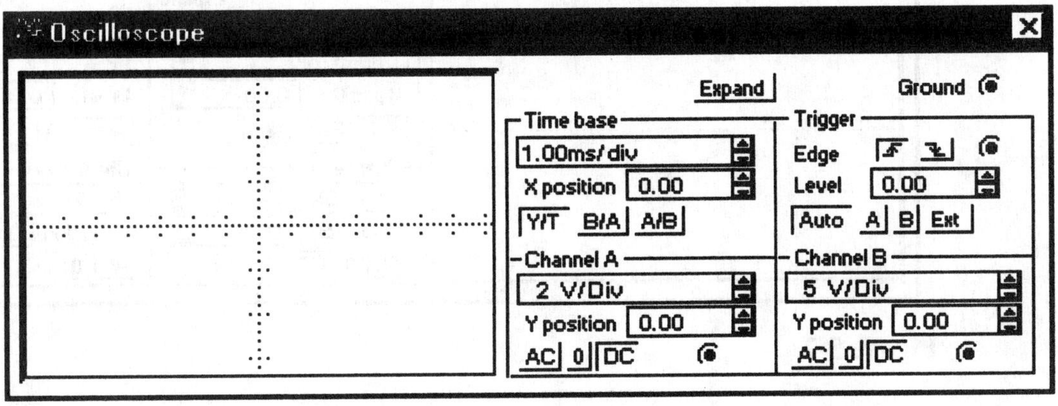

Figure 5-6

Step 4. From your waveform measurement determine the time to ramp up and ramp down. Record.

ramp-up time = _______ ramp down time = _______

Step 5. Calculate the ramp-up and ramp-down times from the key concepts formulas. Show your formulas and solutions here.

Step 6. Leave potentiometer (A) at the same value for a 5 volt peak output. Now adjust the slope that will produce a true sawtooth (the ramp up and ramp down times will be unequal) waveform. Set the slope control for 40% (this sets the non-inverting input to a negative bias voltage) of maximum resistance. Measure the DC voltage (Vy) and record.

Vy = _______

Step 7. Using your calculation (Vy) above, calculate the approximate time to ramp up and the approximate time to ramp down. Show your calculations below.

t(ramp up) = ________ t(ramp down) = ________

Step 8. Draw and label your waveforms on the oscilloscope face below, Figure 5-7.

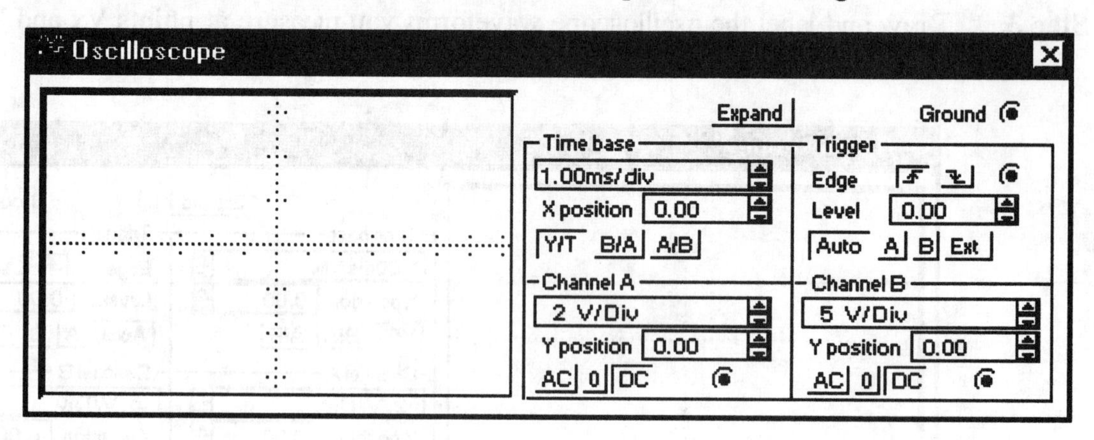

Figure 5-7

Step 9. Set the slope control (S) to 30%. Measure the DC voltage (Vy) and record.

Vy = ________

Step 10. Show your calculations for ramp up time and ramp down time.

t (ramp up) = ________ t (ramp down) = ________

Step 11. Draw and label the waveforms at Vx and Vy on the oscilloscope face in Figure 5-8.

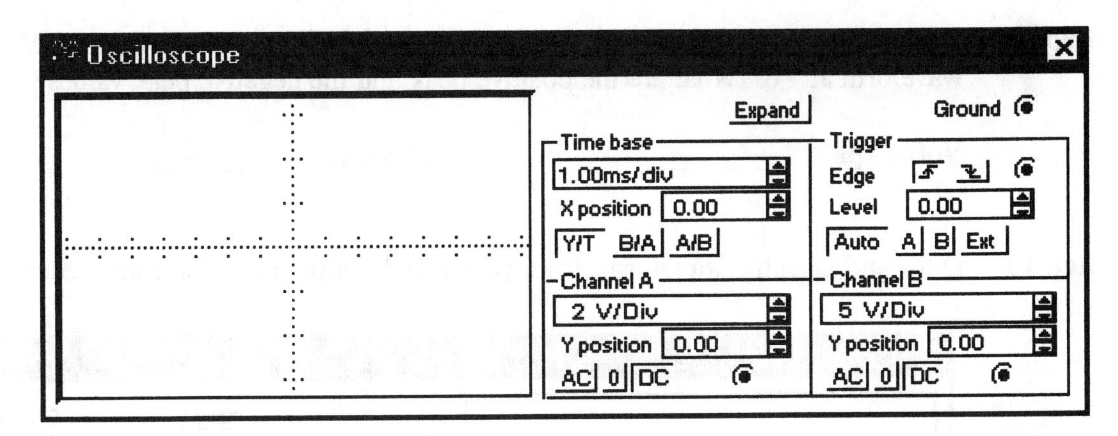

Figure 5-8

Step 12. Return the slope control (S) to 50% and leave the amplitude control (A) for a 5-volt peak output. Change the integrator capacitor to 0.1 uf. Estimate the ramp up and ramp down times by examining the formula and record below.

t (ramp up) = _______ t (ramp down) = _______

Step 13. Measure and record the ramp up and ramp down times for Step 12. They should be close to your estimates.

t (ramp up) = _______ t (ramp down) = _______

Step 14. Draw and label the waveforms Vx and Vy on the oscilloscope face below, Figure 5-9.

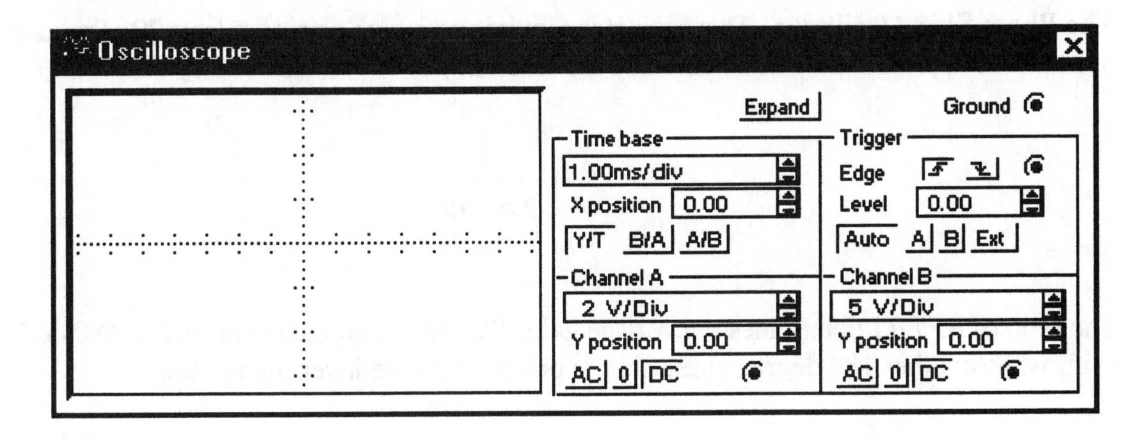

Figure 5-9

Step 15. Remove the diode D1 from the feedback loop of the comparator op-amp. Measure the waveform at Vo. What are the positive peak and the negative peak values?

Vo(pos-pk) = ________ Vo(neg-pk) = ________

Step 16. Draw and label the waveforms from your oscilloscope measurements in Step 15.

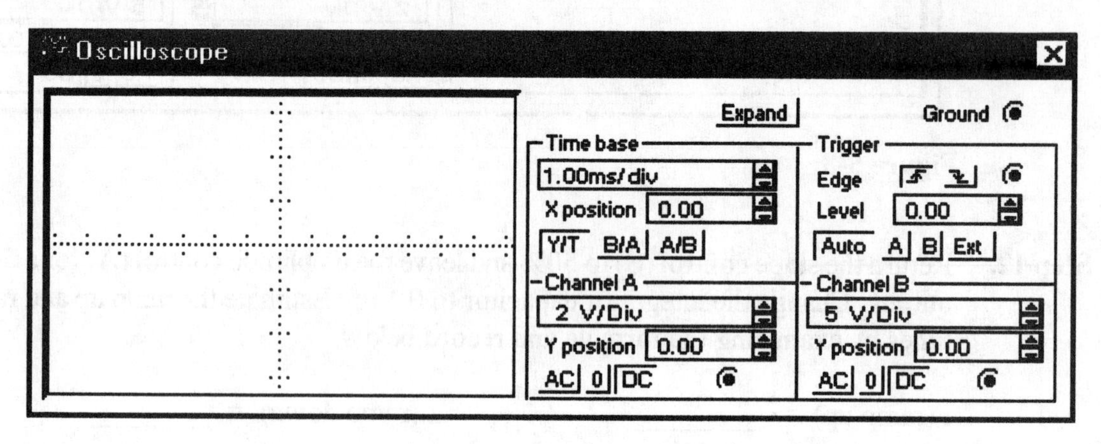

Figure 5-10

Key Concepts
Signal Generating Circuits

The Phase Shift Oscillator. This oscillator requires a voltage gain of at least 29. Frequency of the oscillator is approximately determined by the formula

$$f = \frac{1}{2\pi\sqrt{6}RC}$$

The ratio of RF / R1 determines the voltage gain. The RC components provide a 180 degree phase shift, while another 180 degree phase shift is provided by the inverting op-amp.

Note: If the circuit does not break into oscillation when the ratio of RF / R1 = 29, either increase RF slowly until the circuit oscillates or adjust the oscilloscope vertical input until oscillation begins. This may take a minute or so to occur depending on your computer speed.

The output of the phase shift oscillator is a **sine wave** with no distortion unless the gain exceeds 29, in which case the output peaks will show clipping.

When designing this oscillator choose a common size capacitor and then calculate the value of R from the equation.

The output amplitude can be fixed to a desired peak-to-peak voltage by placing back-to-back zeners from the output of the op-amp to ground.

(RF1 II RF2) / R1 = 29 in Figure 5-11. The diodes turn on to make the circuit self-starting.

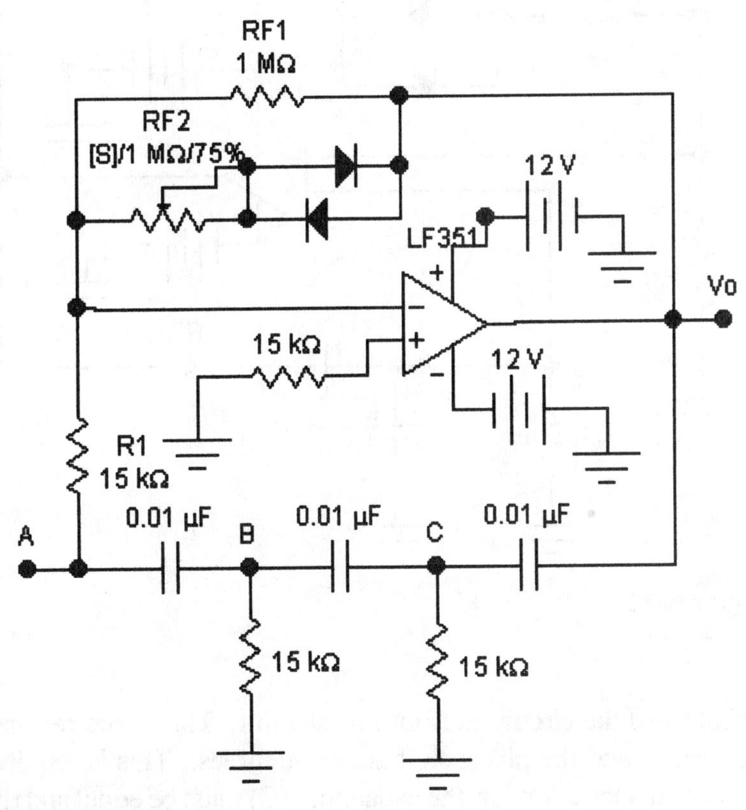

Figure 5-11

The Wien Bridge Oscillator. This oscillator is stable and easy to design at the low-frequency audible range. Generally, a low-value capacitor(s), C, is chosen of 1 uf or less and then resistor (R) is calculated from the following formula:

$$R = \frac{1}{2\pi f C}$$

The top and bottom of the bridge are connected to the inputs of the op-amp, while the sides of the bridge are connected to the op-amp output terminal and ground as shown below, Figure 5-12.

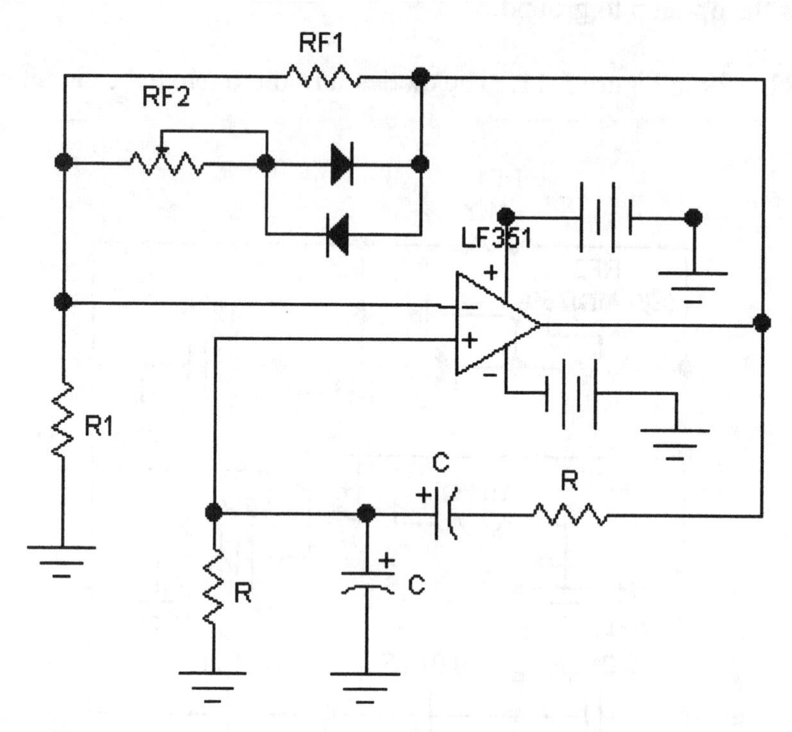

Figure 5-12

When the bridge is balanced the circuit goes into oscillation. This is the resonant frequency of the two RC arms of the bridge and the phase shift is zero degrees. This is required for oscillation to occur in this circuit. For balance to occur, the capacitors (C) must be equal and the resistors (R) must be equal. The series and parallel resonant circuits have the same roll-off frequency, and therefore, the circuit oscillates at the bandpass resonant frequency.

The value of RF, that is, RF1 in parallel with RF2, is determined as follows:

$$A_v = \frac{1}{B} = 3 \quad so \quad 1 + \frac{R_F}{R_1} = 3 \quad and \quad R_F = 2R_1$$

Choose a value of RF1 that will work with a common size potentiometer (rheostat).

The output of this oscillator is a **sine wave.**

The output can be limited in voltage by connecting two back-to-back zeners at the output.

Triangle Wave/Sawtooth Wave Oscillator. The circuit below, Figure 5-13, consists of a zero crossing comparator followed by an integrator.

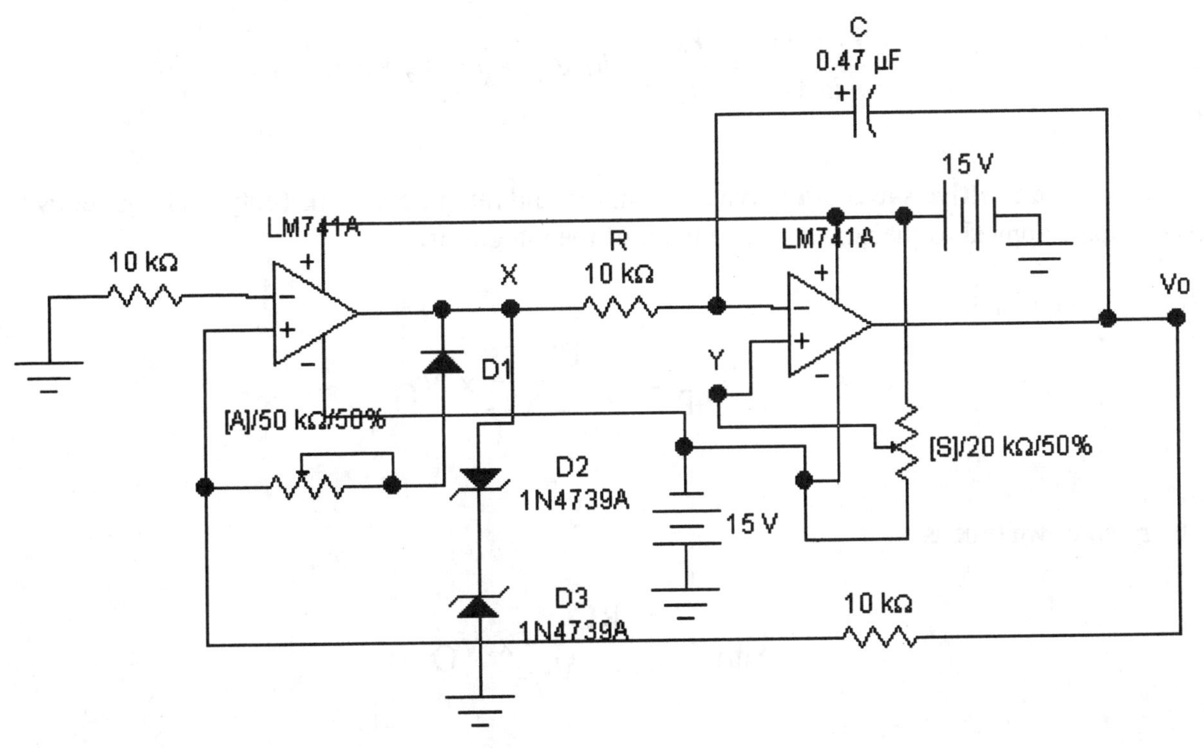

Figure 5-13

With the wiper of potentiometer (S) set at its 50% center position, the signal at the output Vo will be a positive-going triangle wave with equal ramp up and ramp down times. The amplitude of the output voltage of the triangle wave is to be adjusted by the setting of rheostat (A). The peak output voltage of Vo is set by the choice of zener diodes.

With (S) set at 50% and diode D1 not connected in the circuit (shorted out or removed and the rheostat reconnected), the output voltage will be a triangle wave with equal positive and negative peaks of approximately the same amplitude as with the diode connected.

When engineering this circuit for higher frequencies remember that the integrator time constant (T)=RC must be much longer (100X) than the period of the input waveform.

To set the peak output voltage of the sawtooth generator the following formula is used:

Since the current through (A) potentiometer is virtually the same as the current through the feedback resistor to the comparator, then:

$$\frac{V_X - 0.6}{R(A)} = \frac{V_0}{R_F} \quad \textit{where} \quad V_X = V_Z + 0.6v$$

To produce a positive sawtooth wave the ramp-up and ramp-down time (duty cycle) is set by the bias voltage applied to the non-inverting input of the integrator.

The ramp up time is

$$T_{up} = \frac{RC}{V_X - V_Y} \times V_O$$

The ramp down time is

$$T_{dn} = \frac{RC}{V_X - V_Y} \times V_O$$

Chapter 5

Hardware Laboratory -- Design and Build

Design #1

Determine the circuit values needed to design a 2-kHz waveform generator. The output voltage must be a 5 volts peak (positive only) triangle wave.

1. Build this circuit and demonstrate that it works in the presence of your instructor.

2. Prove that the simulation works on EWB by printing the circuit schematic with values of components shown, a parts list, and the working oscilloscope waveforms at Vx and Vy.

Design #2

Design a sawtooth generator that has a 60 percent duty cycle with a 10-volt peak output (positive only). The period of the sawtooth must be no longer than 20 milliseconds.

1. Build this circuit and demonstrate that it works in the presence of your instructor.

2. Prove that your circuit works by printing the circuit schematic with values of components shown, a parts list, and the working oscilloscope waveforms at Vx and Vy.

Questions on Chapter 5
Signal Generators/Oscillators

EWB and Hardware Laboratory Problems are Included

Fill in the correct answers to the questions below.

1. Each RC section of the phase shift oscillator provides _______ degrees of shift.

2. A phase shift oscillator has two feedback paths: one is ____________________ and the other is ____.

3. Generally, the feedback in an oscillator is (positive or negative) _________________.

4. A Wien bridge oscillator uses a balanced bridge, and so oscillation occurs at the ____________________ frequency of the bridge.

5. The triangle wave/sawtooth generator built in this chapter has two sections: one is a _______________________ and the other is a(n) _______________________.

6. The requirement for oscillators to sustain oscillation is that ________________ gain times ____________________ factor must equal _________________.

7. The phase shift oscillator must have a voltage gain of about _________ to sustain oscillation.

8. The Wien bridge oscillator requires a gain of _______ to oscillate.

9. The duty cycle of the sawtooth generator used in this chapter is increased by moving the wiper of the (S) potentiometer toward the _________________ supply voltage.

10. The peak output voltage of the sawtooth generator is increased by moving the ________ rheostat toward ______________________ (less or more) resistance.

CHAPTER 6 Name _______________
TIMING CIRCUITS USED IN DIGITAL ELECTRONICS

INTRODUCTION

The 555 IC timer is a low-cost, general purpose timer that can be used up to about 100 kHz. It is used where a single timing pulse is desired, or where a continuous train of pulses is needed. It is compatible with all digital logic families and with op-amp circuits. The 555 timer has an operating range of about 5-volts to a maximum of 18-volts. The 555 timer IC can be operated in the single-shot or "monostable" mode; which means it outputs a single, positive-going pulse of a duration limited by the practical values of the R and C components that are selected. The output pin returns to zero volts upon the end of timing (T) which is determined by the equation $T=1.1\ RC$.

The second mode of operation is the free-running or "astable" mode. Application of the supply voltage causes this circuit to produce a continuous train of pulses at the output pin. The period of the output pulses is now determined by the equation $T=0.69(RA + 2RB)C$. For further information on the theory of operation and formulas needed for design, turn to the **Key Concepts** at the end of this chapter.

OBJECTIVES

- Understand the operation of the 555 IC timer in the monostable mode.
- Learn to design a monostable timing circuit.
- Understand the operation of the 555 IC timer in the astable mode.
- Learn to design an astable timing circuit.
- Build and test hardware breadboarded timing and interface circuits.

Exercise 6-1, Monostable Timer Circuits

PROCEDURE

Step 1. Load the file labeled **"ex#6-1"** from your EWB data disk. Examine the circuit in Figure 6-1, and note that the trigger input is initially held high resulting in the output pin remaining low prior to a trigger input. Close the switch while observing the output pin with the oscilloscope. Determine the pulse output time. Does this agree with the formula for a monostable timer? Show your calculation and formula below.

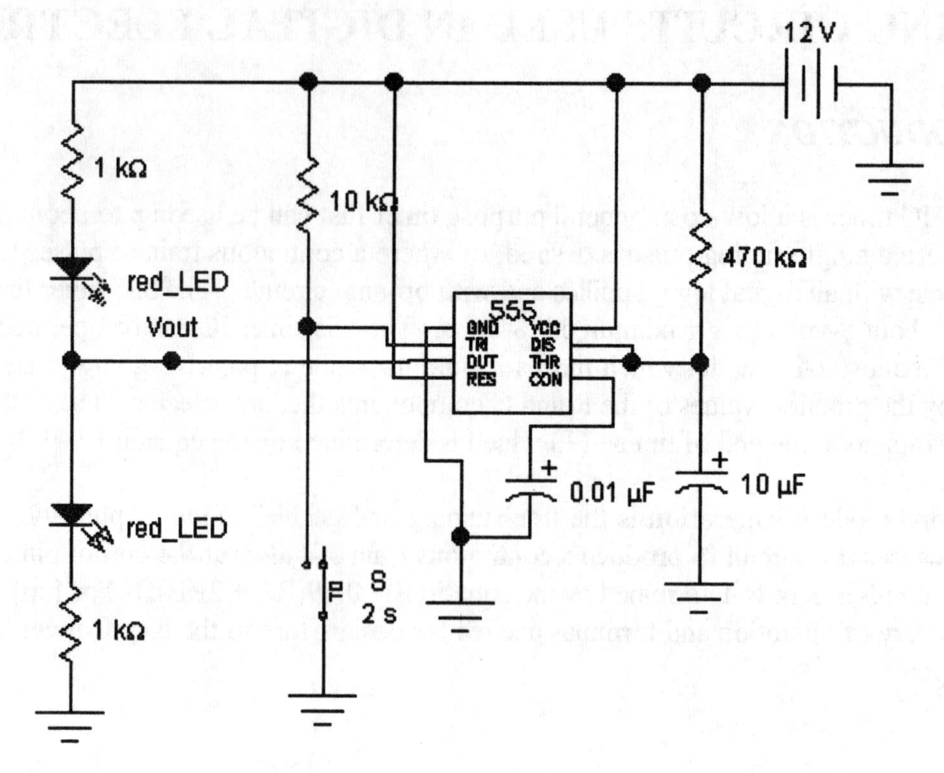

Figure 6-1

Calculation: T(on) = ________

Step 2. At what voltage on the threshold/discharge pins does the timing end?

V = ________

Express the voltage as a fraction of Vcc.

__________ Vcc

Step 3. Disconnect the reset pin from Vcc. Insert a 100 K-ohm resistor from Vcc to the reset pin, then connect a time delay switch (3 seconds on, 4 seconds off) from the reset pin to ground. Record your oscilloscope observation on the oscilloscope face shown below, Figure 6-2. Briefly describe why the timer operates in the manner observed.

Description of circuit operation:

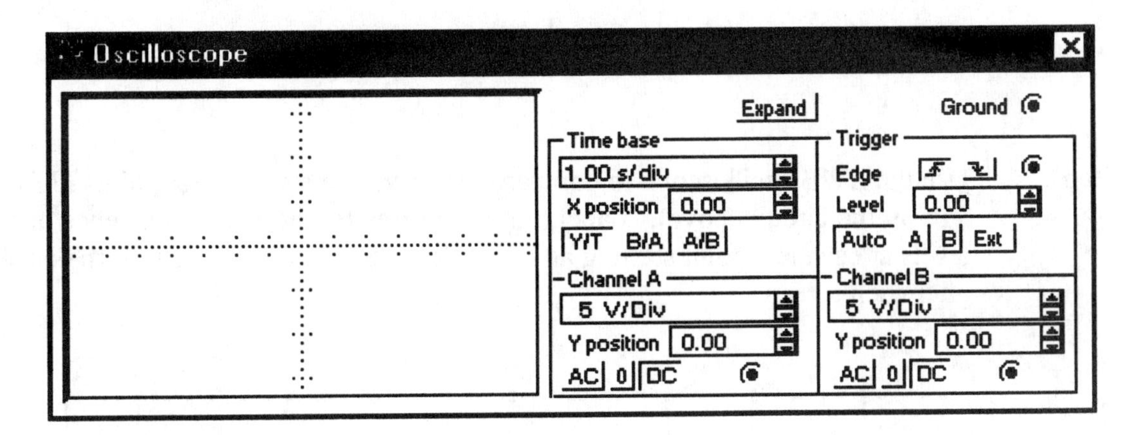

Figure 6-2

Exercise 6-2
Software/Hardware Application of Monostable Timer

INTRODUCTION

Monostable circuit applications are numerous and include missing pulse detection, liquid level tank fill, switch debouncing, touch switch lamp turn-on circuit, and alarm alerting circuits, to name a few. The circuit below, Figure 6-3, is designed to turn on a lamp, pump, or small motor for a specified time interval of 1.1 RC. The input circuit differs from the circuits in Exercise 6-1. The input circuit creates a narrow pulse at the trigger input which allows the input triggering device to actually have a longer pulse time than the timer output pulse time. Consult the **Key Concepts** section of this chapter for the formulas and theory of operation needed to understand this circuit.

Timing Circuits Used in Digital Electronics

Your assignment is to first test the circuit on EWB software. You will be asked to record waveforms and confirm measurements of calculated values at specified test points in the circuit. Next, you will be asked to build and test a hardware version of this circuit to confirm circuit operation and check on measured waveforms and calculations.

Step 1. Modify the previous **EWB** circuit or build a new **EWB** simulation circuit to match the circuit in Figure 6-3. Calculate the amount of time the lamp will be on and record.

T(on) = ______

Step 2. Use the EWB oscilloscope to compare the waveforms at the test points (TP1 through TP5 in the circuit above). Label the waveforms to show time sequence and voltage levels at each test point starting at TP1. Note: Set Analysis to pause the oscilloscope after each screen.

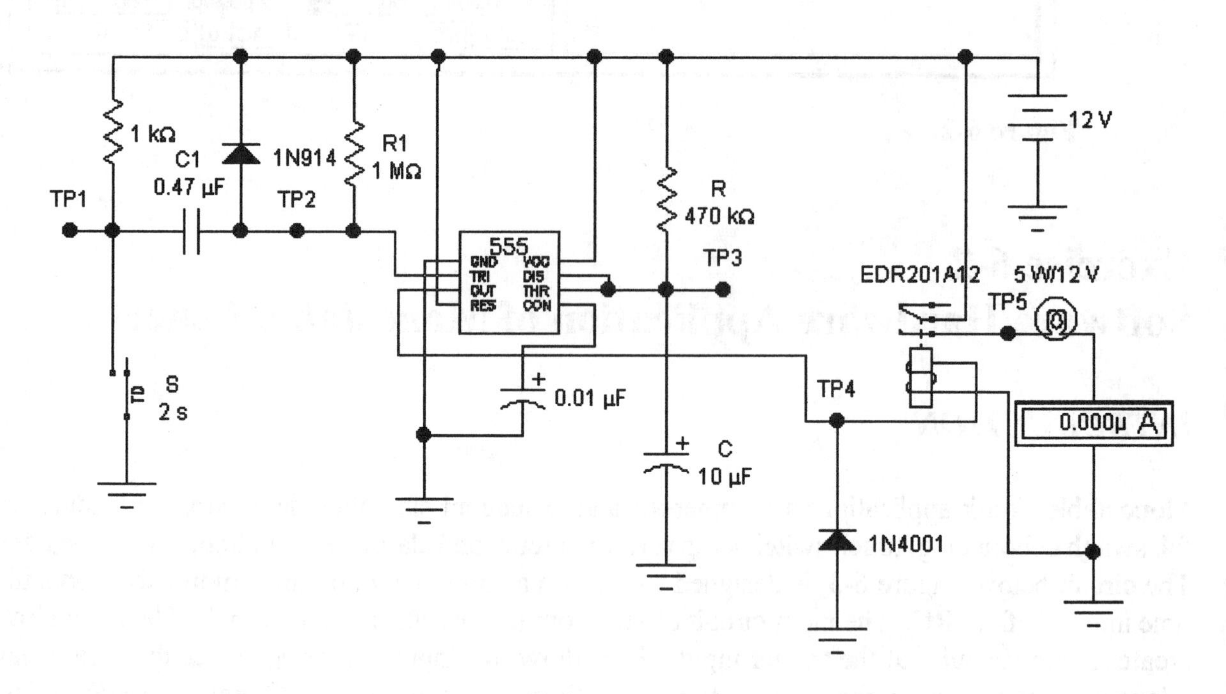

Figure 6-3

Step 3. ***Hardware circuit construction.*** Several considerations need to be examined to make this a functional circuit. The 555 timer IC is rated at 200 mA of output current, but realistically it may not drive the relay that you have selected. If you choose a solid-state relay then you only need to consider the proper voltage levels. However, if you choose a mechanical relay then you must consider the required current to close the relay contacts. If your 12-volt DC relay requires a current of more then 50 mA than you should use a driver transistor to supply the relay closing current.

The second important consideration is the power supply capacity since it must supply a minimum of 500 mA for the 12-volt DC @ 5W light bulb, relay, and 555 circuit. A third consideration is the EWB time delay switch. You may want to use a 556 dual timer to provide the 2-second delay that the EWB switch gave you in the simulation circuit.

Carefully consider each of the above and create a new hardware schematic. Create your new schematic and parts list using EWB software. Print this out to use as a guide in your construction and this will be handed in with the test point waveforms upon successful operation of the circuit.

Step 4. Build and test the circuit in two parts. Test the operation of the timer section first. Do not connect the relay circuit. Use the oscilloscope to measure the voltage level and the amount of time the output remains high at TP4. Record.

Vo = _______ T(high) = _______

Step 5. Test the relay portion of the circuit (no connections to the timer) by applying 12-volts DC to the input of the relay circuit (coil or input terminals on SS relay). Does the lamp light? Measure the DC current required for the lamp. Record.

I (lamp) = _______

Step 6. Connect the timer output to the relay input. Apply power and start the timer sequence. Measure the waveforms at all test points and record the timing sequence with voltage levels as you did in the simulation.

Exercise 6-3, The Astable Timer

INTRODUCTION

The astable multivibrator is self-starting when power is applied to the 555 IC timer. The trigger and threshold pins are connected to the top of the timing capacitor, which means that once timing begins, the capacitor will continuously charge and discharge from 2/3 Vcc to 1/3 Vcc. Since the discharge pin is connected to resistor (RB), the discharge path is through RB only. The charge path for the capacitor includes both RA and RB resistances. Since the trigger pin is at zero volts at initial power-on the output pin goes high and remains high until the capacitor charges to the upper trip point of 2/3 Vcc (threshold voltage). At this point, the capacitor C discharges to the lower trip point of 1/3 Vcc (trigger voltage) and the cycle of charge/discharge repeats.

Duty Cycle is generally defined as the amount of time the output pin is high divided by the period of the output wave. However, it may be defined as low time divided by the period of the output wave. For our purposes, we will use the former definition. Please read the **Key Concepts** section before proceeding to the simulation and hardware applications.

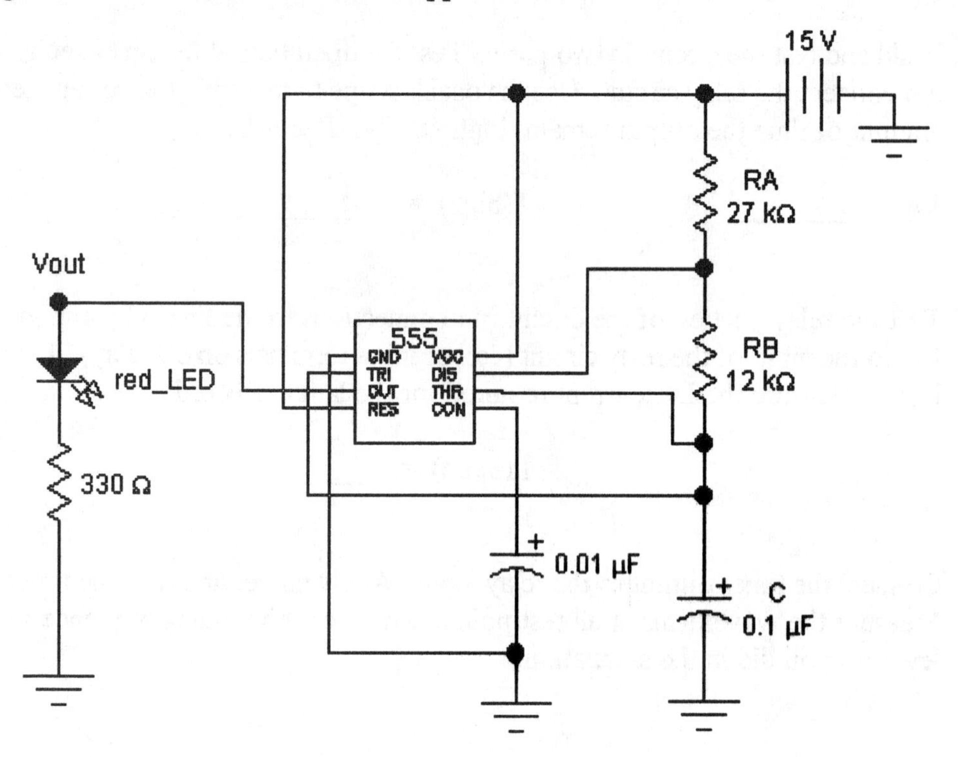

Figure 6-4

Step 1. Load the circuit labeled "**ex#6-3**" from your EWB data disk. Attach CH A of the oscilloscope to the timing capacitor (C), and connect CH B to the output of the timer (pin 3). Set the oscilloscope controls for 5 volts/division and the time base to 0.5 ms/division. Draw the resulting waveforms on the oscilloscope face below, Figure 6-5.

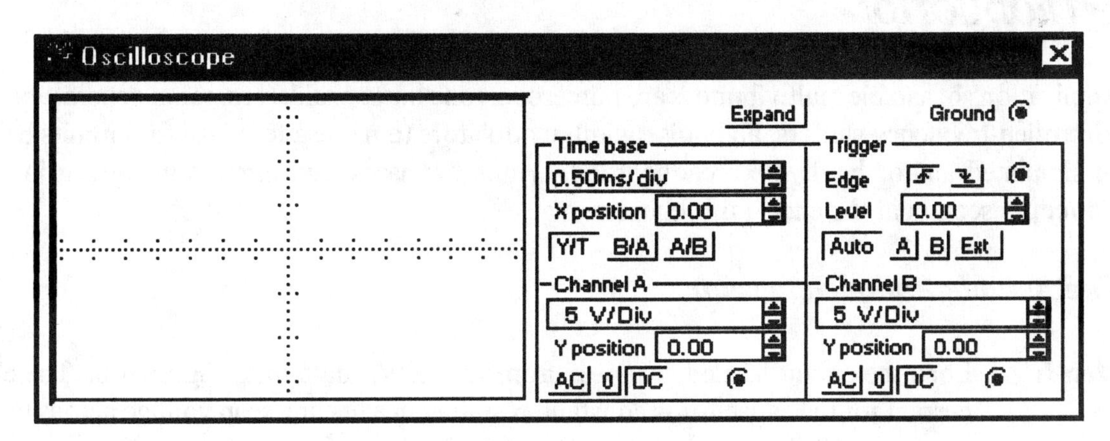

Figure 6-5

Step 2. From your measurements in Step 1 determine the time low, time high, period, and frequency of the output waveform at pin 3.

T(low) = _______ T(high) = _______

T(p) = _______ F(freq) = _______

Step 3. From your measurements, what are the minimum and maximum voltages that appear across the timing capacitor?

V(C) minimum = _______ V(C) maximum = _______

Step 4. From your measurements, calculate the duty cycle (based on time high) of this circuit.

D = _______ %

Step 5. ***Optional Assignment.*** Design and test an astable multivibrator that has a 40% duty cycle and a frequency of 250 hertz. Choose a capacitor of 0.1 uf. Hand in a completed EWB schematic with the components list.

Exercise 6-4, Software/Hardware Astable Ramp Generator

INTRODUCTION

Applications of astable multivibrators are numerous; tone burst oscillators, ramp generators, voltage controlled frequency shifters, and pulse width modulators to name a few. All the circuits below can be simulated and/or hardware breadboarded. Hints for hardwired circuits are given in the **Key Concepts** section at the end of this chapter.

The Astable Ramp Generator

Step 1. Load the circuit labeled "**ex#6-4**" from your EWB data disk, Figure 6-6. The charging current for the capacitor is constant, resulting in a linear rise in voltage across the timing capacitor (C) that ranges from 1/3 Vcc to 2/3 Vcc. Check the ramp voltage across the timing capacitor with CH B while measuring the output voltage at pin 3 with CH A of your oscilloscope. Draw the waveforms and label each waveform on the oscilloscope in Figure 6-7.

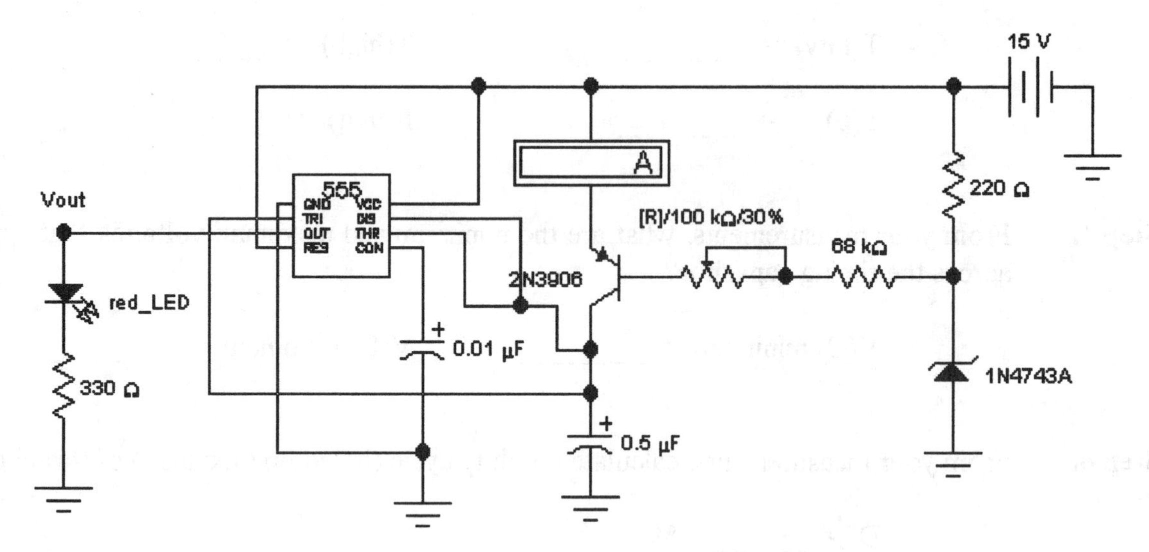

Figure 6-6

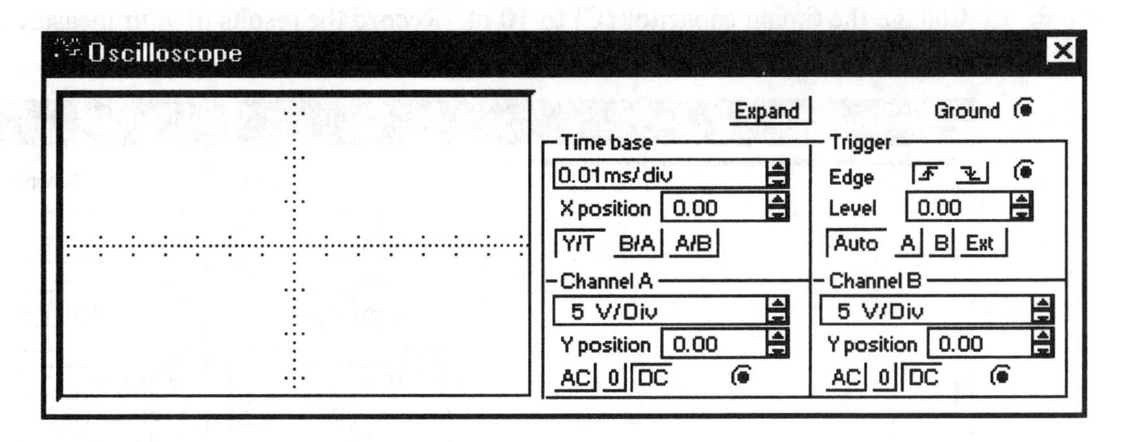

Figure 6-7

Step 2. From your waveform above, determine the time it takes the capacitor (C) to charge from 5 V (1/3 Vcc) to 10 V (2/3 Vcc). Record.

T = _______

Step 3. What is the frequency of the ramp voltage measured in Step 2?

f = _______

Step 4. Change the timing capacitor (C) to 5 uf. Use the oscilloscope to measure the ramp voltage time and frequency. Draw and label the ramp wave on the oscilloscope face below, Figure 6-8.

T = _______ f = _______

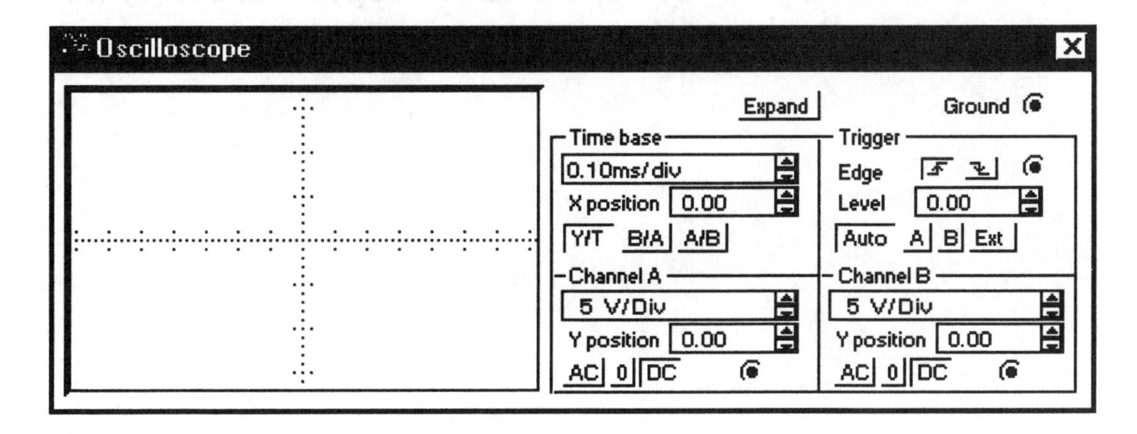

Figure 6-8

Step 5. Change the timing capacitor (C) to 10 uf. Record the results of your measurements as you did in Step 4 above.

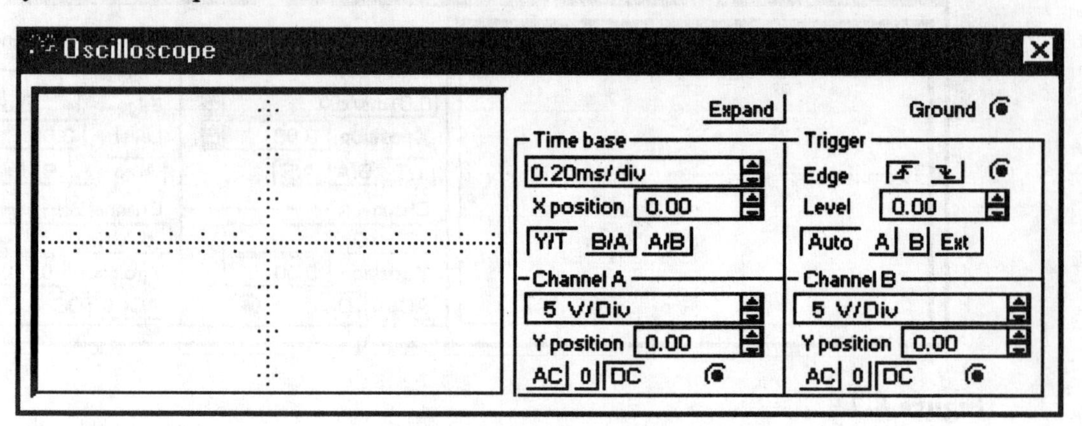

Figure 6-9

T = ________ f = ________

Step 6. Based on your measurements in the steps above, state the relationship between the size of the timing capacitor (C) and the time (T) of the ramping voltage.

Step 7. ***Optional extra credit.*** Develop a formula showing how the time (T) of the ramping voltage can be calculated.

Step 8. ***Hardware Circuit Development*****.** Design a ramp generator circuit using a 555 IC that will ramp from 3-volts to 6-volts with a ramping time of 0.1 seconds and a frequency of 10 hertz. Hint: Use a low-leakage timing capacitor and a base resistor that does not exceed 2.2 M-ohms. Build and test this circuit. Show your schematic (either EWB or neatly drawn in the space below) with component values and specifications.

Step 9. Show your ramp voltage waveform on the oscilloscope face below, Figure 6-10.

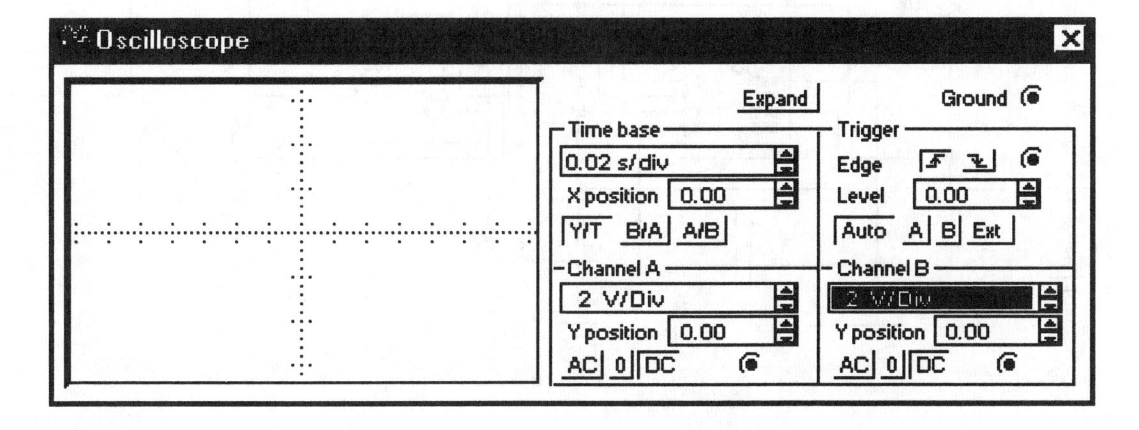

Figure 6-10

Exercise 6-5
Software/Hardware Audible Pulsed Alarm Circuit

INTRODUCTION

The circuit below, Figure 6-11, makes use of the fact that the reset pin can be used to enable or disable a 555 astable timer. Remember that when the reset pin is below one volt this overrides all other conditions, and pulls the output down to logic zero. Load the file "**ex#6-5**."

In this circuit disabling the first astable timer also disables the second astable timer. When the switch is moved (using the space bar) from the ground position, the reset input rises to Vcc which starts both astable timers. The first astable timer starts its longer timing period while the second astable timer pulses the buzzer during this time period. The first astable timer goes low, which then shuts off the second timer (because its reset goes to 0 volts) until a new timing cycle of the first astable timer begins.

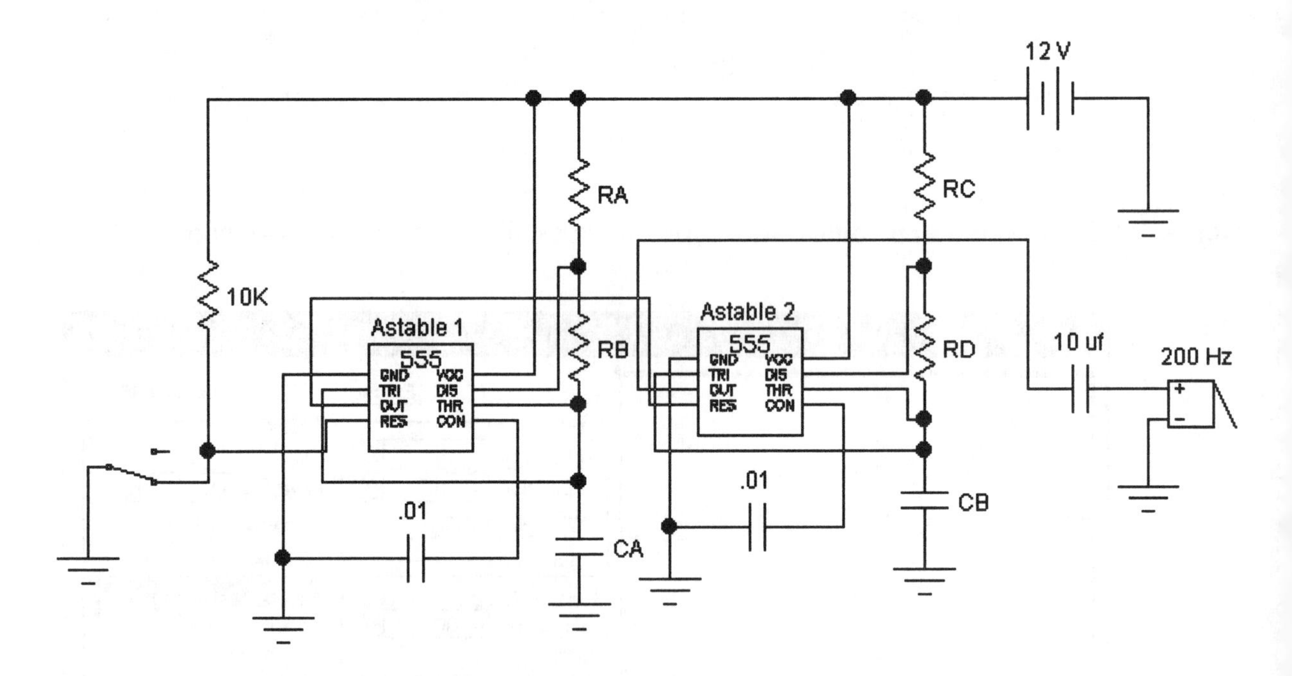

Figure 6-11

Step 1. Make resistors RA and RC each 1 K-ohm, RB and RD each 1 M-ohm, CA = 5 uf, and CB = 0.5 uf. Calculate the periods of the astable 1 and astable 2 timers.

T (1) = _______ T (2) = _______

Step 2. Connect oscilloscope CH A to the output of astable timer 1 and CH B to the output of astable timer 2. Set the X position of CH A to 0.2 for clear viewing of the output of astable timer 1.

Record the waveforms on the oscilloscope face below, Figure 6-12.

How many pulses are delivered to the speaker during the on time of the astable timers? Why?

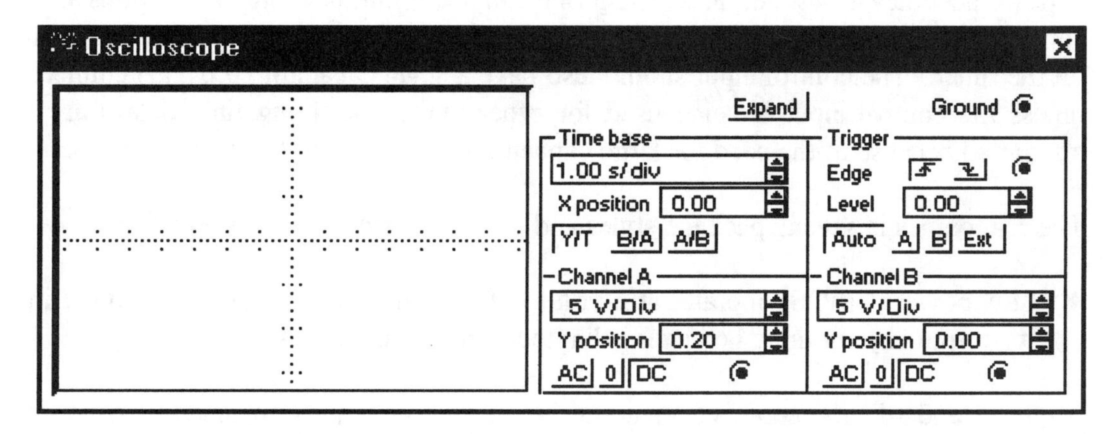

Figure 6-12

Step 3. ***Hardware circuit development.*** Design and build an Audible Pulsed Alarm Circuit that will create three pulses at the buzzer when astable timer 1 is held high for three seconds. Each astable timer should have an approximate 50% duty cycle. Make sure that your buzzer has an adequate driving source. Turn in a schematic with its parts list and demonstrate to your instructor that the circuit works properly. Place your schematic in the space below.

Key Concepts
Timing Circuits Used in Digital Electronics

The 555 IC timer is a versatile timer used in many general purpose applications where pulses are needed, a ramping voltage is required, a rectangular wave with a specific duty cycle is required, or in many other applications where timing or counting is required. It is low in cost and is compatible with systems where 5-volt to 18-volt (DC) voltage is used in analog and digital electronics.

The 555 IC timer is limited to less than 100 kHz in frequency and lacks the stability needed when extreme accuracy is desired. False triggering can occur unless wiring precautions are used. A filter capacitor of up to 100 uf should be connected near the point where Vcc connects to the power pin on the timer. The control input should also have a small capacitor (0.01 uf) connected to ground unless the control input is being used for other purposes. Long time delays are generally not attempted because of the need for large capacitors which can be leaky and unstable.

The 555 comes in several package styles and several versions to meet user needs.

- Low power (CMOS) operates at 2-volt to 18-volt and consumes only about 0.1 mA at 15-volt. Its output current drive however is limited to less than 20 mA.

- The standard 555 operates from about 5-volt to 18-volt and consumes almost 15 ma at 15-volt. It is rated at 600 mW, but typically will supply only about 40 mA of drive current.

Monostable Circuit Operation

The timing period for completion of a single output pulse is initiated from a high to low transition at the trigger input. This "falling" voltage must fall below one-third of the power supply voltage for the output to be triggered high through the internal lower comparator and bistable flip-flop (see the simplified circuit below). The lower comparator sets the bistable flip-flop, and the output terminal goes to a high state (near Vcc).

The capacitor (C) then charges through the resistor (R) until the voltage at the threshold terminal reaches 2/3 (Vcc). At this point, the bistable is reset and Q' goes high, driving the internal transistor into conduction which causes the external capacitor to discharge, taking the output of the 555 to a low state (near zero volts).

The period (T) to the output pulse is $T = 1.1\ RC$.

The width of the trigger pulse should be less than one-fourth of the period (T).

t (time) < 1/4(T)

If the reset pin is taken low (below one volt) timing stops and the output goes low.

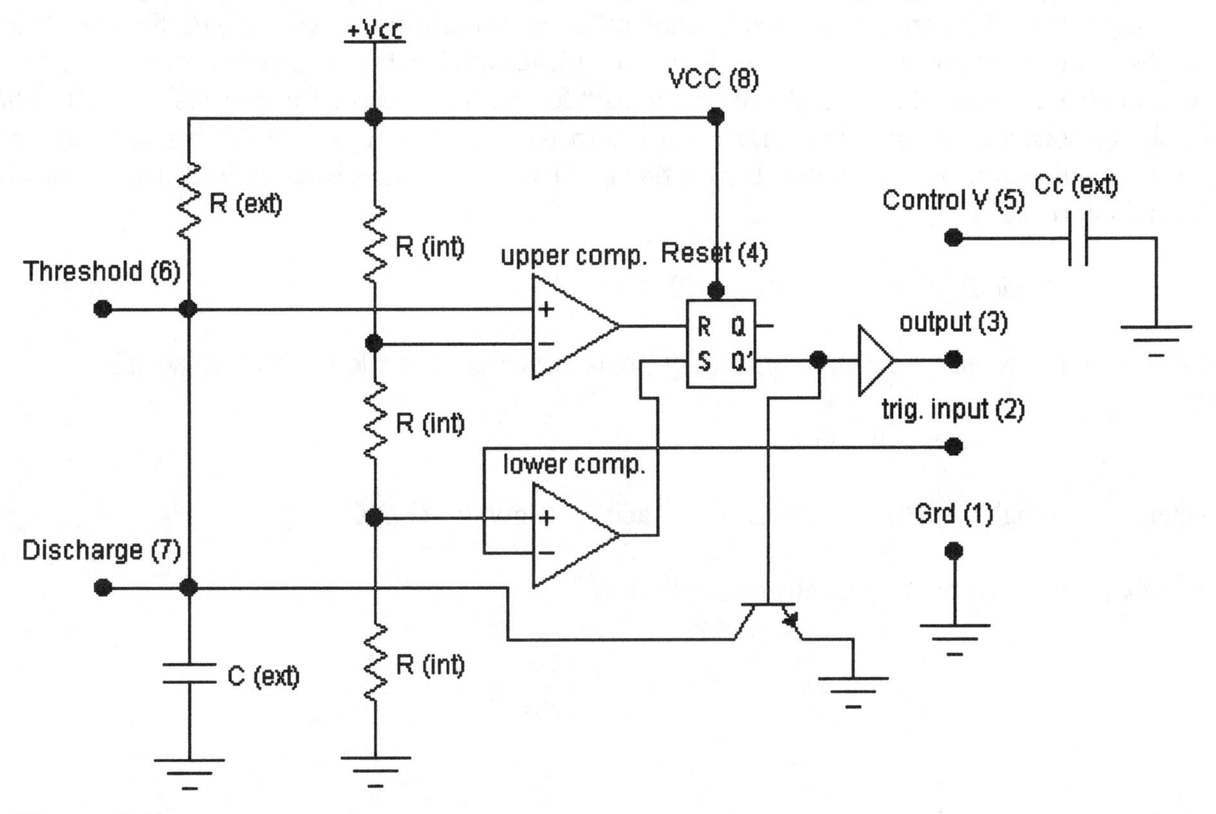

Figure 6-13

Monostable Design Example

Given: Vcc = 12-volt

Problem: An output pulse of 100 msec is needed

Solution:
1) Select a timing capacitor (C). Arbitrarily choose 0.1 uf
2) Choose control capacitor (Cc) to be 0.01 uf
3) Solve for R from the formula T = 1.1RC
   R = 90.9 K-ohm (use a 100 K-ohm potentiometer)
4) Choose a trigger input pulse of less than one fourth of (T). Choose a trigger pulse width of 10 msec. Arbitrarily choose C(t) = 0.22 uf and solve for R(t) from T(t) = R(t) × C(t)

   R(t) = 39 K-ohm (approximately)

Note: Exercise 6-2 in this chapter uses the trigger pulse narrowing circuit used in Step 4 of the solution above.

Astable Circuit Operation

Note the two major modifications the astable multivibrator has compared to the monostable circuit. First, the trigger input and threshold input are both connected to the timing capacitor. Second, the discharge terminal is connected to the junction of RA and an additional resistor RB. So, the circuit is self-starting once power is turned on. Now our charge path for the timing capacitor includes both resistors RA and RB. However, the discharge path for the timing capacitor includes only RB. The discharge occurs when the upper comparator is turned on by the fact that 2/3Vcc has been reached when enough charge has accumulated on the timing capacitor. The time is determined mathematically from the formula:

$$\text{(period) } T = 0.693\,(RA + 2RB)C$$

The period (T) consists of charge time t(high) plus discharge time t(low). Mathematically:

$$T = t(high) + t(low)$$

where, $t(high) = (RA + RB)C$ and $t(low) = (RB)C$

The frequency (f) is mathematically determined by

$$f = \frac{1}{T} \quad or \quad f = \frac{1.44}{(RA + 2RB)C}$$

Duty cycle (D) can be considered as time high divided by the period to the waveform (time high plus time low). This is the general case in most texts, although duty cycle could be time low divided by the period of the waveform. The former expressed mathematically as percent duty cycle is:

$$\%\text{duty cycle} = \frac{t(high)}{T} \times 100$$

Astable Design Example

Given: Vcc = 12-volt

Problem: Design an astable circuit for frequency (sometimes called pulse rate frequency p.r.f.) of 1 kHz with a duty cycle of 80%.

Solution:

1. The period of the pulse waveform as shown above is the reciprocal of frequency, or as shown below:

$$T = \frac{1}{1 \times 10^3} = 1 \text{ ms}$$

and t(high) equals 80% of the period (T), so t(high) equals 0.8 ms.

2. Therefore, time low is the remaining time in the period or 0.2 ms. So:

$$RB = \frac{t(\text{low})}{0.693(C)} \approx 29 \text{ K} - \text{ohm}$$

3. RA plus RB is calculated from:

$$RA + RB = \frac{t(\text{high})}{0.693(C)} \approx 115 \text{ K} - \text{ohm}$$

So RA equals 88 K-ohms

The circuit is shown below, Figure 6-14, with the resulting waveforms (Figure 6-15) across the capacitor and at the output.

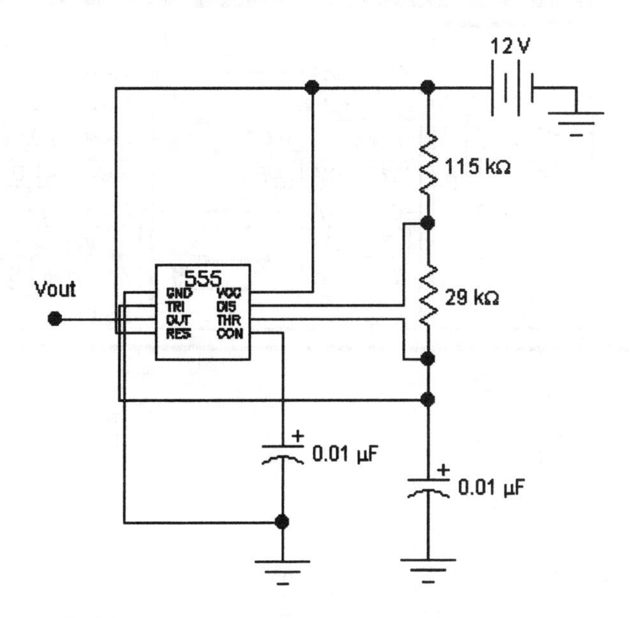

Figure 6-14

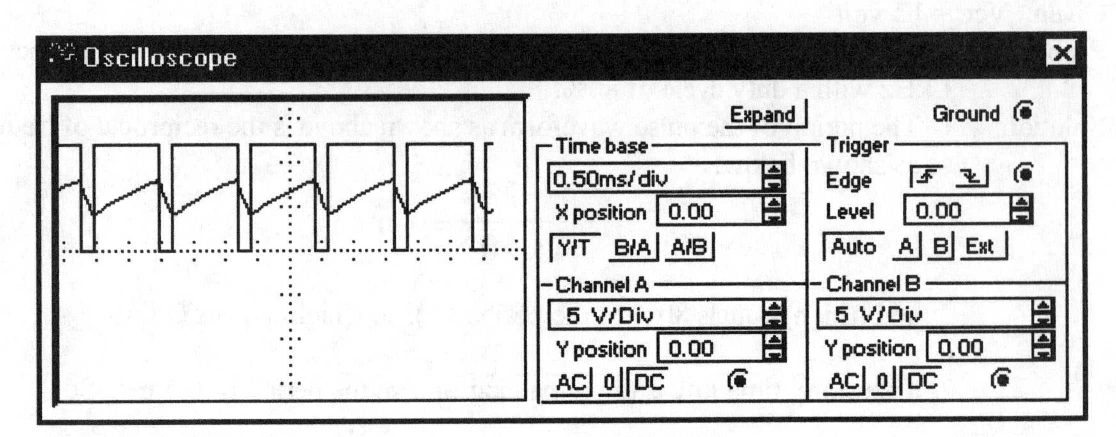

Figure 6-15

Astable Ramp Generator Operation

The astable ramp generator produces a linear rising voltage that begins at 1/3Vcc and continues until it reaches 2/3Vcc, at which point it falls to 1/3Vcc in just a small fraction of the rise time. The waveform then repeats itself as all oscillators do. The linear rise in voltage is produced by a constant current flowing to a capacitor. Notice the difference in the capacitor waveform from the oscilloscope waveform shown above.

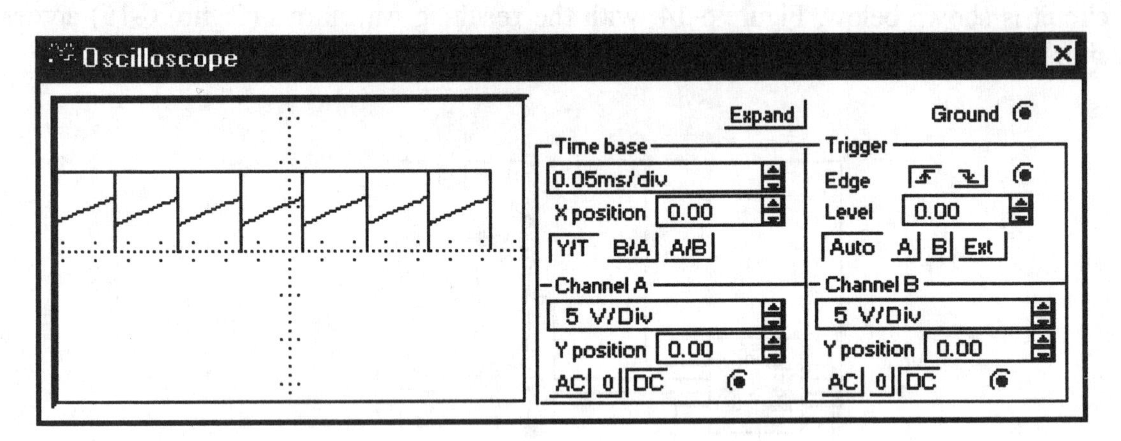

Figure 6-16

Astable Ramp Design Example

Problem: You need a linear ramp from 3-volts to 6-volts and the ramp-up time must be 100 usec.

Solution: Since the ramp-down time will be very short the frequency of the astable waveform will be 10 kHz.

1. Use a power supply of 9-volt DC. Arbitrarily choose a common capacitor size of 0.1 uf. The additional charge on the capacitor needed for the voltage rise of 3-volts (1/3Vcc to 2/3Vcc) is determined by the formula:

$$Q = CE = 0.1 \text{ uf} \times dv\,(3v) = 0.3 \text{ uC}$$

The current the capacitor needs to accumulate this charge is:

$$Q = I \times T \quad \text{so} \qquad I = 0.3 \text{ uC} / 100 \text{ us} = 3 \text{ mA}$$

The constant current circuit to develop 3 mA is shown below.

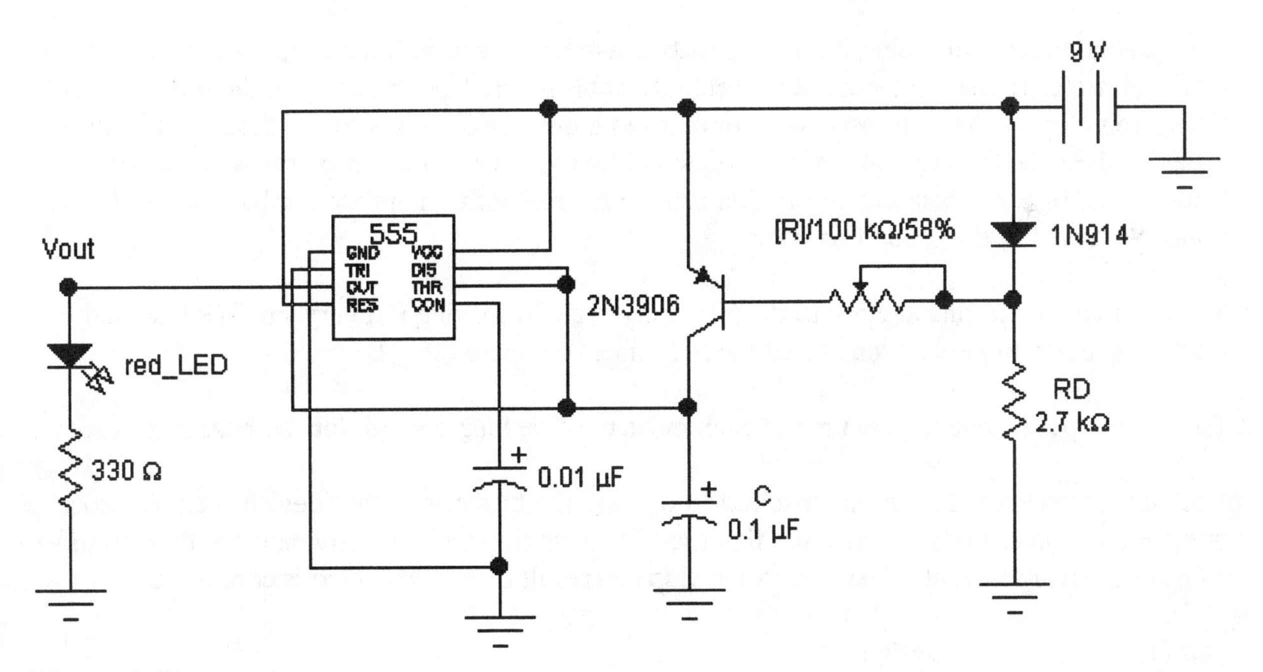

Figure 6-17

2. If a 6.2-volt zener is used, allow about 10 mA of zener current.

$$RD = \frac{(9v - 6.8v)}{10\,mA} = 220\ K - ohm$$

3. The base resistance needed will be approximately:

$$R_B = \frac{(V_B - V_Z)}{I_B} \quad \text{Where:} \quad IB \cong \frac{3mA}{Beta} \approx \frac{3mA}{180} = 16.6\ uA$$

Then:

$$RB = \frac{8.3\,v - 6.2\,v}{16.6\,uA} \cong 125\ K - ohm$$

Choose a potentiometer of 100 K and a series resistor of 100K .

4. Adjust the potentiometer until the ramp-up timer is 100 us.

Audible Pulsed Alarm Operation

The astable pulsed alarm circuit consists of two astable timers with the output of the first astable timer connected to the reset input of the second astable timer. Upon opening of the spdt switch both timers begin operation. The second timer delivers a number of pulses to the buzzer while its reset is held high by the first astable timer. The second timer ceases to deliver pulses when its reset input is taken low by the timing out of the first astable timer. Both timers are disabled when the input switch is moved to the ground position.

The timers are operating at close to the 50% duty cycle by making RA very small (RA should be at least 1000 ohms to prevent loading of the 555 timer) compared to RB.

To hear the pulsed buzzer, the time of each pulse must be long enough for the buzzer to respond.

You must determine the current required to operate the buzzer or other device such as a speaker prior to any connections. Remember that the 555 timer chip will not provide more than 40 mA to 50 mA of driving current. Damage to the chip can result if excessive load is connected.

Audible Pulsed Alarm Design

Problem: Deliver four pulses to a buzzer each time the output of the first astable timer goes high. The time high for the first astable timer is 10 seconds.

Solution: Since this is a relatively simple problem it is left up to the student to solve and test on EWB.

Questions Chapter 6
Timing Circuits Using The 555

Simulation and Hardware Laboratory Questions

True or False (Monostable Timers)

______ 1. The monostable starts its timing when the trigger input goes from a high to a voltage less than Vcc/3.

______ 2. The monostable timer is considered to be an oscillator.

______ 3. If the output is high, a ground on the reset pin immediately forces the output low.

______ 4. R1 and C1 in Exercise 6-2 create a narrow trigger pulse for the input of the 555 timer.

______ 5. Diode D1 in Exercise 6-2 limits the voltage at the trigger input to 12-volts plus the diode forward voltage.

______ 6. The pulse width at the output is determined mathematically to be 1.1 times the width of the trigger pulse.

______ 7. The voltage across the timing capacitor (C) rises from zero to 2/3 (Vcc).

______ 8. The time (T) of the output pulse of the 555 monostable is 1.1 (RC).

______ 9. Without the diode (1N4001) in Exercise 6-2, the 555 could malfunction or be damaged by the collapsing field of the relay.

______ 10. The 555 is fully capable of driving the lamp directly in Exercise 6-2.

True or False (Astable Timers)

______ 1. Duty cycle of the astable timer can be defined as the output time high divided by the period (T).

______ 2. The charging waveform of voltage across the timing capacitor (C) is linear.

______ 3. The voltage across the timing capacitor ranges from 1/3 (Vcc) to 2/3 (Vcc).

______ 4. It is not possible to build an astable multivibrator with a 50% duty cycle.

______ 5. A ramp generator requires a constant current supplied to the capacitor (C) during its charge time.

______ 6. Doubling the size of the timing capacitor will double the voltage at the output terminal of the 555.

______ 7. The ramp voltage is a linear rising waveform that appears across the timing capacitor (C).

______ 8. The timing capacitor discharges through an internal transistor in the 555 when the output is low.

______ 9. The period (T) of an astable multivibrator is determined by the timing circuit components RA, RB, C and the power supply voltage.

______ 10. The frequency of the ramp generator in this lab is almost equal to the reciprocal of the ramp-up time.

CHAPTER 7

Name ________________

DIGITAL TO ANALOG & ANALOG TO DIGITAL CONVERTER PRINCIPLES AND APPLICATIONS

INTRODUCTION

The importance of D to A (DAC) and A to D (ADC) conversions to electronics is tremendous. They are an essential part of any instrumentation system and are being used extensively in modern consumer electronics equipment. DAC's and ADC's are often built into microprocessor/computer systems. Transducers that measure light, heat, sound, pressure, moisture, magnetic fields, and other similar things are analog devices. A microprocessor system understands only digital 1's and 0's. Therefore, analog to digital conversion is necessary if one wishes to gather (input) and interpret transducer information. Likewise, for the microprocessor system to send signals (output) to analog devices such as small DC motors, valves, and servo amplifiers, a digital to analog conversion must take place.

In this chapter you will learn the basic concepts of each converter and learn the key terms and what to consider in choosing a converter. Since microprocessor systems are generally used with these devices, you will also learn the necessary connections and timing considerations for this interface to work correctly.

OBJECTIVES

- ♦ Learn the principles of an R-2R ladder as a simple digital to analog converter.
- ♦ Convert a current DAC to a voltage output.
- ♦ Gain an understanding of resolution and input-output equations.
- ♦ Simulate ADC and DAC operation.
- ♦ Hardware wire ADC and DAC chips to demonstrate a practical application.
- ♦ Develop an understanding of connecting converters to microprocessor systems.

Exercise 7-1, The R-2R Ladder

PROCEDURE

Step 1. Read the operation theory of the R-2R ladder in the **Key Concepts** section at the end of this chapter. Load the file labeled "**ex#7-1**" from your EWB files disk. Make sure that all switches are in the actual ground position. Connect an ammeter and measure the current entering the ladder from Vref and record.

I(ref) = ________

Step 2. Connect an ammeter between the i (out) connector and the inverting input of the op-amp. Close switch (0) as shown in the schematic below, Figure 7-1. Measure and record the current going to the op-amp.

i (out) = ________

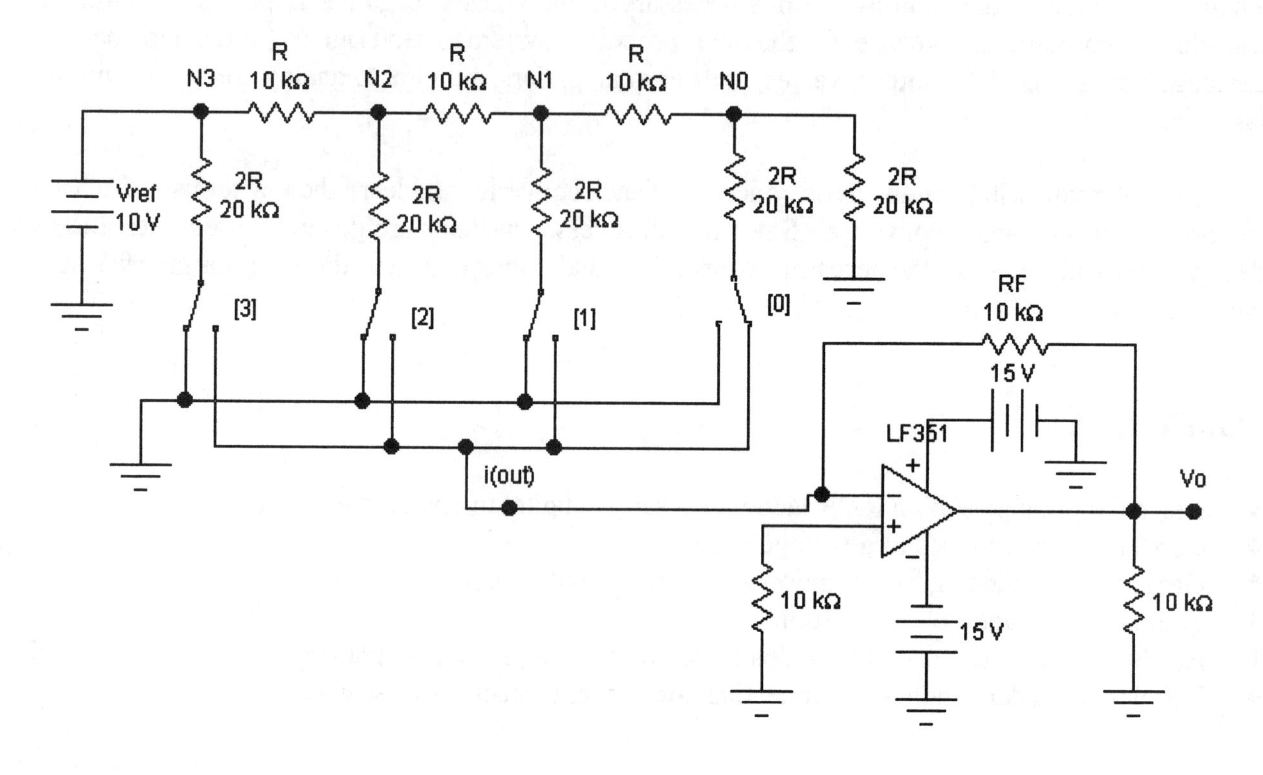

Figure 7-1

Step 3. Do your calculations for I (ref) and i (out) agree with your measurements? Show your calculations below.

Step 4. Measure and record the voltage at Vo with only switch (0) connected to i (out). Show your calculation of Vo below.

Vo = ________

Step 5. Close switches D(0) and D(1). Measure and record the current i(out). Measure and record the new output voltage. How much larger is the current in the D(1) path compared to the current in the D(0) path?

i (out) = ________ Vo = ___________

Step 6. Close the remaining switches D(2) and D(3). Measure and record the current i(out) and the output voltage Vo.

i (out) = ________ Vo = ___________

Step 7. Leave all resistors in this circuit the same value. Calculate the value of Vref necessary to get a −10 volts to appear at the output. Show your calculation below. Now adjust Vref to make Vo equal to −10 volts. Simulate the circuit to see if your calculation is correct.

V(ref) = ________

Step 8. Remove Vref from the circuit and attach the function generator in its place. Use an appropriate coupling capacitor to connect the generator to the ladder circuit. Set the function generator to 10 volts peak to peak at a frequency of 1 kHz. With the oscilloscope, determine the maximum output voltage of the op-amp if only switch D(0) is closed.

V(max) = ________

Step 9. What is the peak output voltage you can attain in this circuit with all the switches closed?

V(max) = ________

Exercise 7-2
IC DAC-08 Simulation and Hardware Application

INTRODUCTION

The DAC-08 is a multiplying digital to analog converter integrated circuit using the R-2R ladder internally. This DAC is capable of operating at high speeds and produces a current output of several milliamperes; typically 2 mA to 3 mA. You will examine the operation of the DAC-08 in conjunction with a voltage amplifier op-amp in the unipolar (+5 volt) and also in the bipolar output mode (+5 volt to −5 volt). The DAC as a digitally controlled volume control that can be used in an application with or without a microprocessor system will be simulated and discussed as a possible hardware project. Read the DAC section of the **Key Concepts** section at the end of this chapter prior to beginning this exercise.

Step 1. Load the file labeled "**ex#7-2a**" from your EWB data disk. The schematic is shown below, Figure 7-2. Write the formula to find the reference current for this DAC. Insert an ammeter and measure the reference current. Record.

Formula for I(ref) _______________ Measured I(ref) = _______

Step 2. Move all bit switches except the bit (0) switch to the ground position. Record the current leaving the DAC to the op-amp inverting input.

i (out) = _______

Step 3. Predict the output voltage of the data switches D7 ----- DO (7-O) for the following conditions:

a) 0000 1111 Vo = _______

b) 1000 0000 Vo = _______

c) 1111 1111 Vo = _______

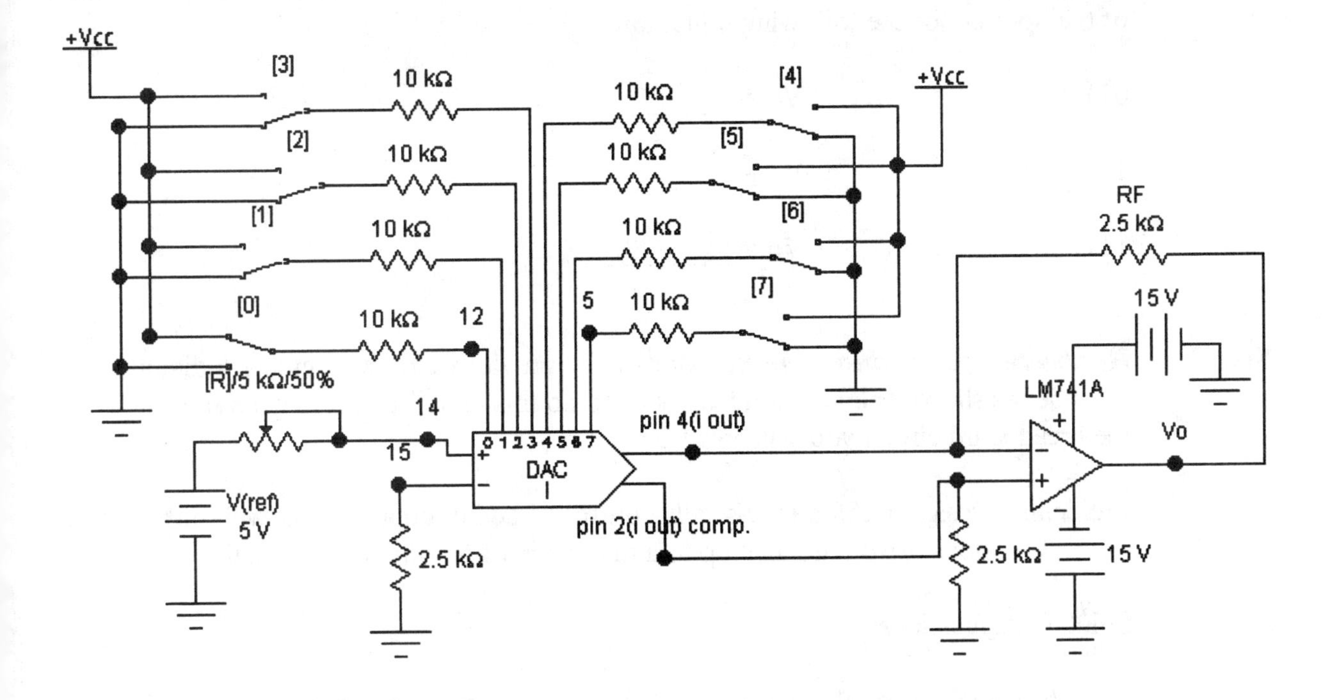

Figure 7-2

Step 4. Load the file labeled "**ex#7-2b**" from your EWB data disk. This circuit is set up for a bipolar output that has a range of approximately positive 5 volts to negative 5 volts. Set switch (0) high and set all the remaining switches low. Measure the current in the i(out) line and measure the current in the i(out) complement line and record each below.

i (out) = ________ i (out) complement = ______

Step 5. From the formulas in the **Key Concepts** section of this chapter, calculate the voltage output based on your current readings in Step 4. Show your calculation below.

Vo = ________

Step 6. Using the ammeter readings as you did in the previous steps, determine the output voltage of the op-amp for the following conditions:

0111 1111 Vo = _______

1000 0000 Vo = _______

1111 1111 Vo = _______

Step 7. ***Hardware circuit construction.*** Modify the previous circuit to create a digital DAC amplifier as shown below. You may want to do this on EWB just to get a schematic for the breadboard circuit you will construct.

Problem: Design a 256-step digitally controlled audio amplifier that will output equal steps of voltage gain up to a maximum of 5 volts peak-to-peak.

Solution suggestions:

1. The signal you apply to the (+Vref) input must never go negative, so you must use an offset if using the function generator.

2. The digital input values can be supplied by an 8-switch DIP package using 5 volt for the Vcc value. A microprocessor system with an appropriate interface system to output ones and zeros could also be used.

3. Use the oscilloscope to measure the gain of the amplifier at several steps. For example, the 0000 0000 digital input should give you zero volts at Vo. If not, you need to use an offset circuit on the op-amp to set the zero level. (See **Key Concepts**). 0100 0000 will give you one-quarter of maximum i(out) and therefore will result in one-quarter of the maximum output voltage.

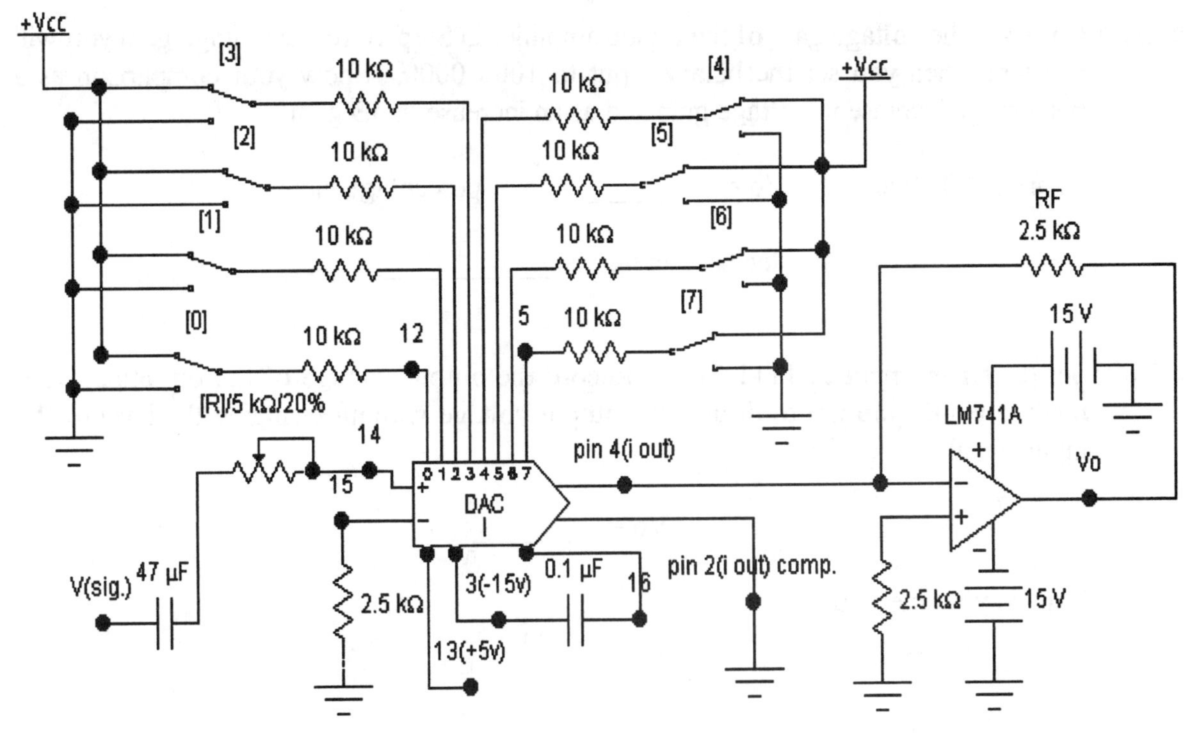

Figure 7-3

Step 8. Calculate the peak input voltage signal required to produce an op-amp output signal of about 2.5 volts peak. Note: V(sig.) equals Vref.

Peak Vref = ________

Step 9. Using your calculation in Step 8, now calculate the peak current entering pin 14 of the DAC. Record this current and also the LSB current that would leave pin 4 of the DAC.

Peak i(ref) = ________

Step 10. Set the binary input value to 0100 0000 and measure the op-amp output voltage. Record. If this voltage is not one-fourth of the required peak-to-peak output voltage, you need to go back and analyze the theory, then redo the steps.

Vo = ________

Step 11. Compare the voltage gain of the digital amplifier in Step 10 to the voltage gain you will measure when you set the binary input to 1000 0000. Show your comparison as a percentage increase in voltage gain and as an increase in dB gain.

binary 1000 0000 Vo = _______ percentage increase is _________

decibel increase is __________

Step 12. Set the binary input to 1111 1111. Record the output voltage of the op-amp. What method would you use to double the output voltage without changing the level of the input signal?

Vo = _______

Explain your method:

Exercise 7-3
ADC Simulation and Application

INTRODUCTION

There are several methods of converting an analog DC input voltage to a digital output value. The ADC 0801 uses the successive approximation method. Two analog inputs are provided to allow differential input measurements (Vin + and Vin−). The output is 8 bits of binary that can readily be used for analysis by a microcomputer. The internal clock (you must select an external R and C value) generates the timing pulses needed for conversion. The outputs are tri-state latches to hold information until a new conversion is requested. Read the **Key Concepts** at the end of this chapter for important information on status and timing considerations for proper operation of this A to D converter chip. Additional applications of A to D converters will be presented in Chapter 10.

Step 1. Load the circuit labeled "**ex#7-3**" from your EWB data disk. The 555 is an adjustable duty cycle timer used to provide a start of conversion signal to the ADC IC chip. Examine the circuit below, Figure 7-4, and record the connections of Vref− and OE. Also note the voltage of Vref+.

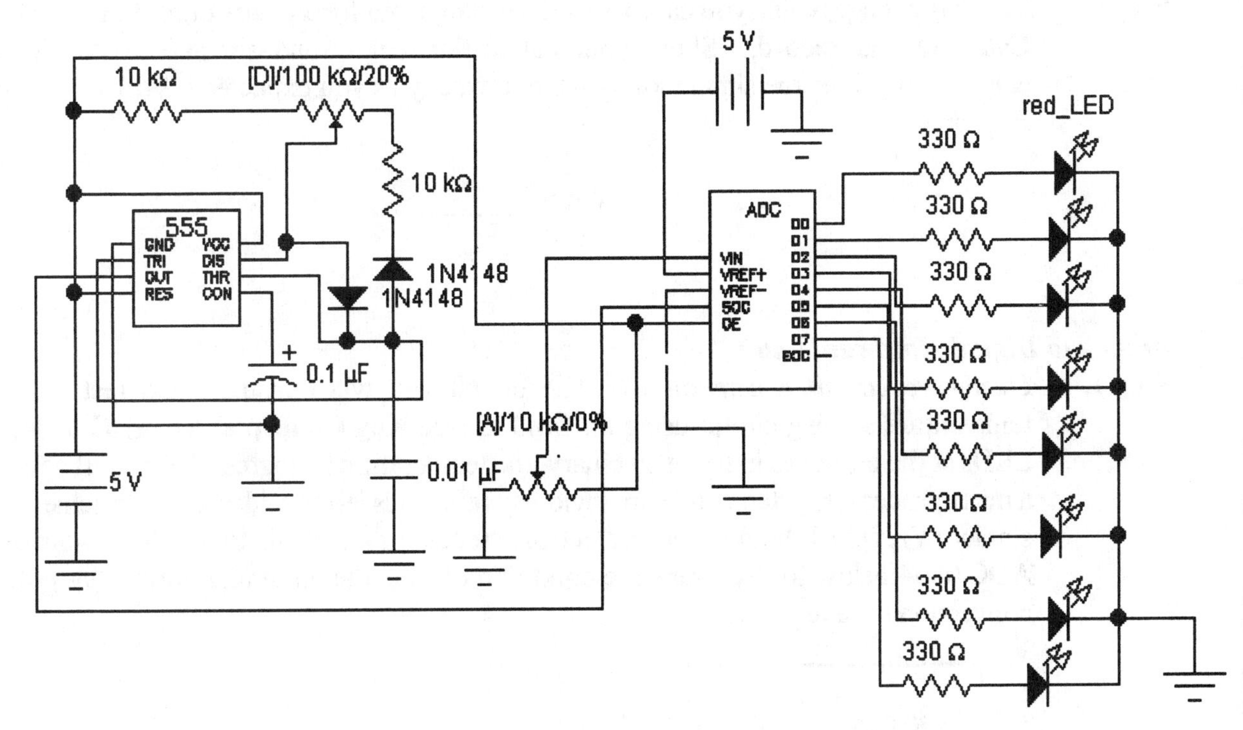

Figure 7-4

Step 2. Set the duty cycle (D) of the 555 timer to 50%. Set the analog input potentiometer (A) to 1%. Record the frequency and duty cycle that is measured. Also record the LEDs that are lit and the DC analog voltage you measure at Vin.

Note: You cannot light just the least significant LED (0) because 1% is the lowest percentage allowable on EWB.

f = ________ Duty cycle D = ________%

Vin = ________ LED's on ____________ (0 is LSB)

Step 3. Calculate the voltage you would need at the analog input pin to light only LED (0).

Vin = ________

Step 4. To what voltage would you have to set the analog input for a binary output of 1000 0000. Calculate and record. Show your calculation below and simulate to verify your calculation. Does the simulation come out exactly as you expected? Why or why not? Explain.

Vin = ________

Analog to Digital Hardware Lab

Step 1. ***Hardware circuit construction.*** The circuit you will construct and test will be a temperature sensing circuit using an LM335 (see **Key Concepts**), an ADC 0804, and LEDs at the output to indicate, in binary, the temperature in degrees Celsius. If you have a microprocessor system, you can obviously refine this lab considerably. To achieve best stability Vref pin 9 should have an electronic regulator attached. Study the pin-out of the ADC 0804 below to determine the condition of each the input and output lines for the conversion to take place.

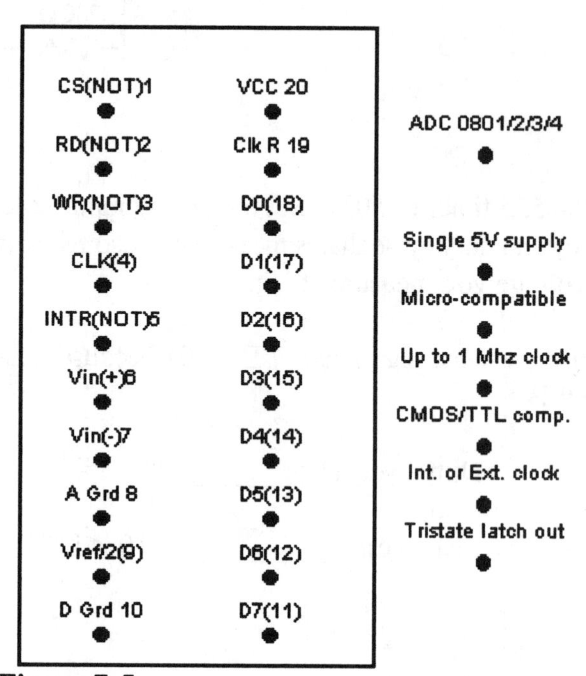

Figure 7-5

Step 2. Wire the temperature sensor circuit as instructed in the **Key Concepts** section of ADC conversion. Connect this circuit to the Vin(+) pin (6) of the ADC.

Step 3. Attach either an electronic regulator circuit or the Vref potentiometer to the Vref/2 pin (9) of the ADC.

Step 4. Attach a 555 astable timer (or a microprocessor system with the appropriate program) to the start of conversion pin. This is the WR (NOT) pin (3) on the ADC.

Step 5. Complete the wiring. A block diagram of the circuit is shown below, Figure 7-6.

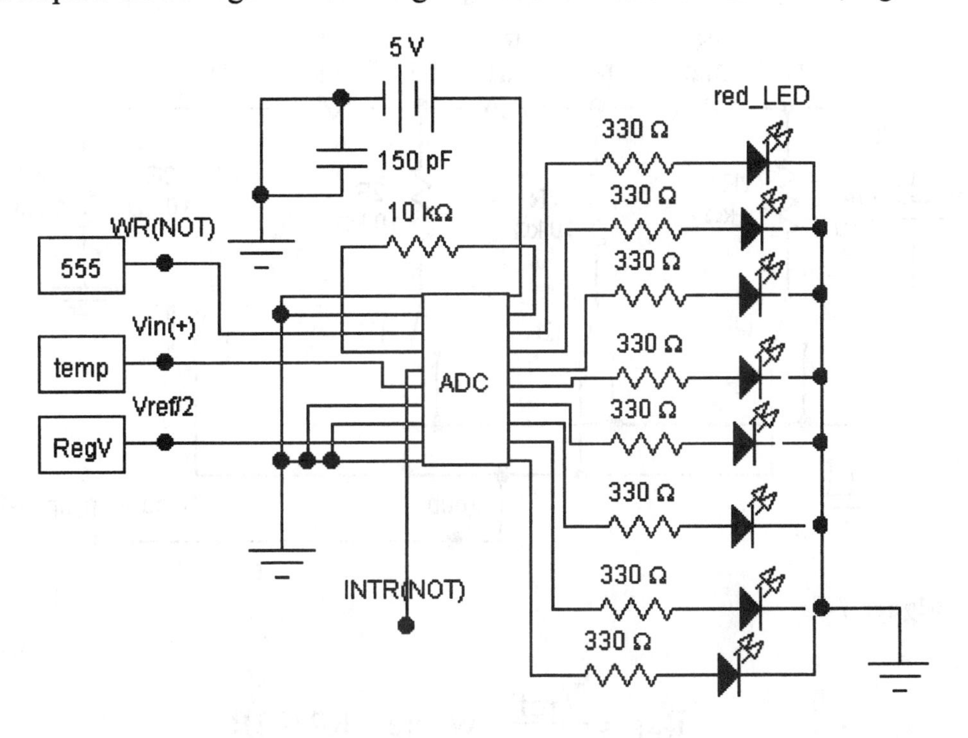

Figure 7-6

Step 6. Calibrate the temperature sensing circuit by placing the LM335 in ice to set to zero Celsius, and then either checking against room temperature or attaching the temperature sensor to a surface of known temperature.

Step 7. What is the binary value that appears on your LEDs if the room temperature is 22 degrees Celsius?

T(binary) = _______________

Key Concepts
Digital to Analog Conversion

A resistive DAC uses an R-2R ladder network consisting of only two resistor sizes. The resistance from any node to ground is equal to R in the circuit below, Figure 7-7. Keep in mind that the inverting input of the op-amp is a virtual ground. The current from the voltage source Vref is always equal to Vref divided by the resistance R. Examples are shown below the figure.

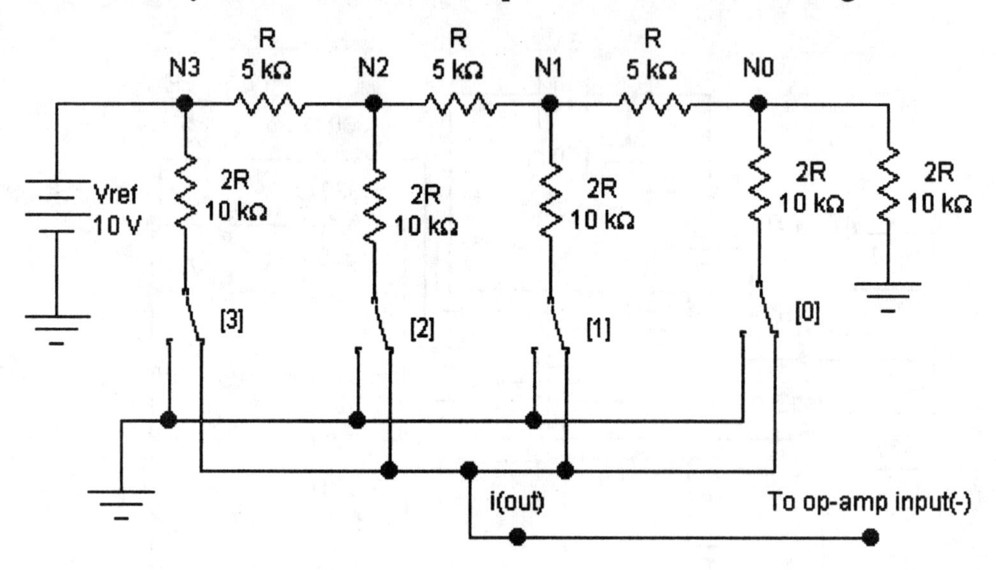

Figure 7-7

$$I_{ref} = \frac{V_{ref}}{R} \quad \text{where} \quad R2 \, II \, 2R$$

$$I(O) = \frac{I_{ref}}{2N} \quad \text{and} \quad i(out) = I(O) \times D$$

So the value of current to the op-amp equals the least significant bit current (resolution) multiplied by the decimal value (D) of the digital input word.

If you connect the ladder current output to the inverting input of an op-amp there will be a voltage conversion and the voltage output of the op-amp will equal:

$$V_o = -i(out) \times RF$$

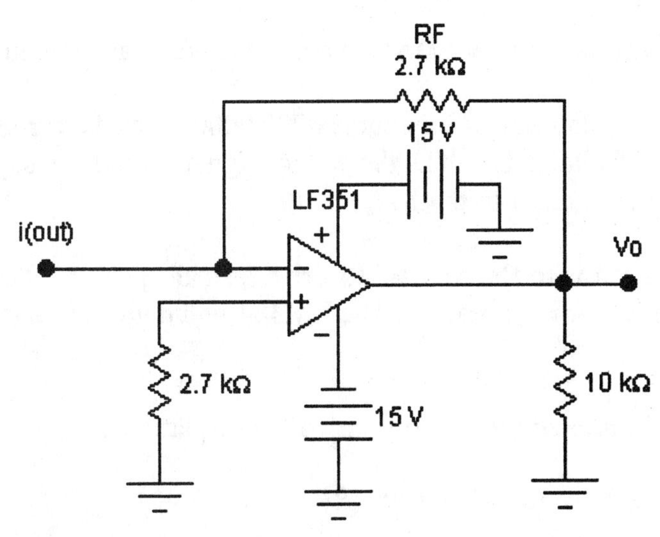

Figure 7-8

The DAC-08 (LM 1408, MC 1408) is a fast MDAC (multiplying D to A) converter with pin layout as shown below, Figure 7-9.

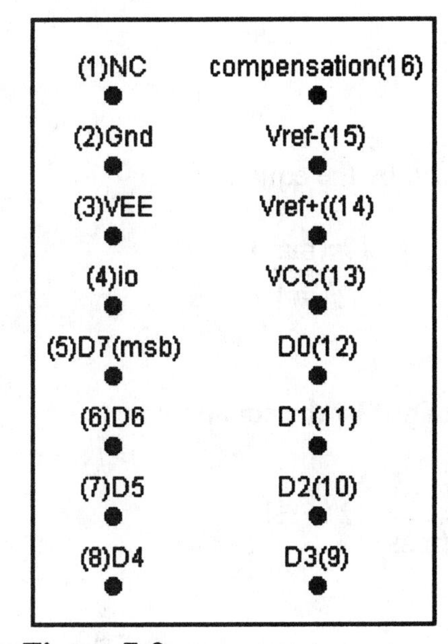

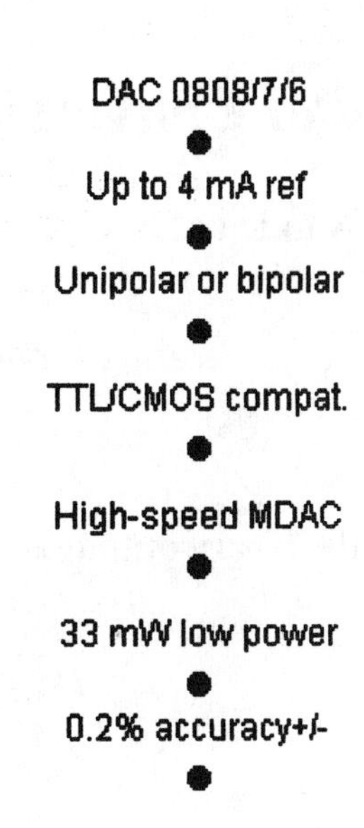

Figure 7-9

The resolution of the DAC is expressed by the equation:

$$resolution = \frac{Vo(\max)}{2^n - 1}$$

The DAC 0808 can be connected as a **multiplying DAC** in a digital audio gain configuration.

The digital input could be set to any of 256 ranges (of course 0000 0000 would be zero volume). A microprocessor would supply the digital bits while the audio signal would be supplied by a preamplifier. The audio signal would connect to pin 14.

A small power amplifier could be used with the MDAC to drive a small speaker. The LM380 is capable of 2.5 watts(rms) to drive an 8-ohm speaker. The LM384 will deliver 5 watts and is pin compatible.

When the DAC 0808 is wired for ***bipolar output*** the output voltage equation is:

$$V_o = [i(out) - \overline{i(out)}]RF$$

Therefore, a 1 LSB increase actually results in double the current increase. This means that the bipolar connection will result in twice the output voltage of the unipolar connection.

Analog to Digital Conversion

The resolution of the A to D (ADC) is determined by the equation:

$$resolution = \frac{Vin(\max)}{2^N - 1}$$

The decimal value of the binary output is determined by the equation:

$$D = \frac{Vin}{Vin(\max)}[2^N - 1]$$

The ADC 0808 is a successive approximation converter; a very fast A to D converter. You can try to observe the conversion time by looking at the end of conversion pin (EOC). It is only a few microseconds and can be used for digitizing audio signals. Internally it has a DAC, a comparator, and a successive approximation register. Each bit is looked at successively until a final bit pattern is determined.

A dual slope A to D converter is a much slower type used in applications such as DMM's.

The flash converter is the fastest, and is used in video digitizing. Cost increases with resolution.

The ADC 0808 is microprocessor system compatible. Other types such as the AD 7574 and AD 670 are excellent choices for microcomputer interfacing.

The microprocessor-compatible ADC must have the following attributes:

1. a tristate memory register to hold the last conversion.
2. a high impedance state to the data bus when the chip is not used.
3. a chip select (CS) input to connect or disconnect the chip (also called chip enable (CE)).
4. Read and Write pins or a single R/W pin. (Write is usually taken low when doing a conversion.)
5. an INTR pin or end of conversion to let the microprocessor know when a conversion is completed.

Hardware Lab Operation Using the ADC 0808

The 555 timer provides the start of conversion signal. The write pin (WR or SOC) must go from a high state to a low state for conversion to begin. Also note that the chip select line (CS) must be in a low state for the chip to be ready for use. The read pin (RD) must also be connected low for conversion to start. The INTR pin goes active low when the conversion is completed. Connecting the INTR pin to the WR pin will result in continuous conversion.

If using a microcomputer, it is important that the proper sequence of 1s and 0s be given and read from the analog to digital converter. The computer must be programmed to give the proper sequence. A simple program would be:

1. send out bits from port(s) to select chip and activate start of conversion.
2. check end of conversion pin and let microprocessor know if conversion is done.
3. bring the digital byte into the computer.
4. tell the ADC to end conversion by sending out proper bits to pins, or tell the ADC to continue looking for a conversion.

Connection of the LM 335 or LM 35 Temperature Sensor.

The **LM 335** temperature sensor requires an external current-limiting resistor and a calibration potentiometer. The calibration potentiometer should be adjusted at some known temperature, Figure 7-10.

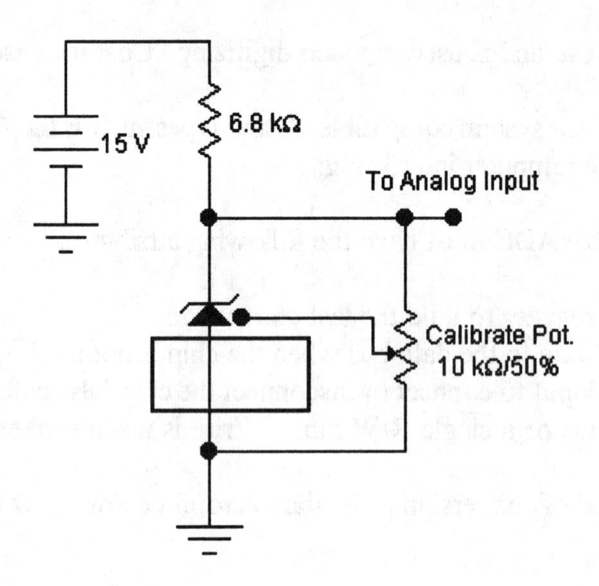

Figure 7-10

The voltage drop across the LM 335 is found by multiplying the temperature in Kelvin by 10 mV. The temperature range is from 233 degrees K to 373 degrees K (−40 to 100°C).

The **LM 34** is a Fahrenheit sensor, while the **LM 35** is a Celsius temperature sensor. Both of these will operate from 4-volt to 20-volt, down to a temperature of about 2 degrees. They put out 10 mV per each respective degree increase in temperature and require no external components except a power source to operate.

Chapter 7 Questions
D/A & A/D Principles and Applications

Simulation and Hardware Questions

Digital to Analog Conversion. True/False.

_____ 1. A 10-bit DAC has a final output voltage of 10-volts. The resolution of this DAC is 10 mV/1 LSB.

_____ 2. An 8-bit DAC has a final output voltage of 10-volts. A binary input of 0100 0000 will have an output voltage of 2.5-volts.

_____ 3. The user of a DAC-08 can adjust the reference current to a maximum of 4 mA.

_____ 4. The resolution of any converter depends only on the number of bits used.

_____ 5. In a multiplying DAC (MDAC) the output voltage is the product of two input signals.

_____ 6. If the chip select pin is high on an ADC 0808 converter then conversion will occur.

_____ 7. A successive approximation ADC is the fastest of all the A to D converters.

8. A DAC-08 has a reference voltage of 5-volts and a reference resistor of 2.5 K-ohms. It is attached to an op-amp with a feedback resistor of 5 K-ohms. The resolution (mV/bit) is:

____

9. What is the output voltage of the op-amp for the circuit of problem 4 if the binary input is 1000 0000?

____

10. A 4-bit binary ladder has rail resistors of 5 K-ohms and rung resistors of 10 K-ohms. The resistance that Vref always "sees" is:
    A. 10 K-ohms
    B. 5 K-ohms
    C. infinite
    D. depends on RF only

Converter Principles and Applications

11. The input voltage (considered full scale) of an ADC 0804 converter is 2.55 volts when a binary output of 1111 1111 appears at the output. Find the resolution of the ADC.

Resolution __________

12. In problem #8, find the binary output when the input voltage is 0.64 volts.

Binary output ______________

13. An 8-bit analog to digital converter has a resolution of 20 mV per LSB. What is the quantization error?

Q(error) ______________

14. Name two types of analog transducers.

A.

B.

15. What is meant by terminals labeled SOC and EOC on an ADC?

CHAPTER 8
POWER SUPPLY CIRCUITS

Name ________________

INTRODUCTION

Integrated circuit regulators are used to supply power to many modern low-power circuits where only low currents from milliamperes to an ampere or two are required. In this chapter you will start with a full-wave capacitor-filtered power supply and proceed to examine the circuits within a typical electronic regulator. You will also learn to build a regulated power supply and boost the current output of a regulator by adding external components.

OBJECTIVES

- Determine the performance of a full-wave, unregulated power supply.
- Examine the operation and performance of an op-amp regulator circuit.
- Build a full-wave, regulated power supply and test its operating characteristics.
- Design, simulate, and analyze a bipolar power supply.
- Calculate and select the components needed for constructing a regulated power supply.
- Learn measurement techniques to evaluate a power supply.

Exercise 8-1, Full-Wave Unregulated Power Supply

PROCEDURE

Step 1. Load the file labeled "**ex#8-1a**" from your EWB data disk. Read the **Key Concepts** section at the end of this chapter so that you may do the calculations specified. Measure the DC voltage that appears across the filter capacitor with no load attached. Measure the RMS voltage at the secondary of the transformer. Record these values and the DC current indicated by the ammeter.

V(DC) = _______ V   V(sec) = _______ V

I (load) = _______ mA

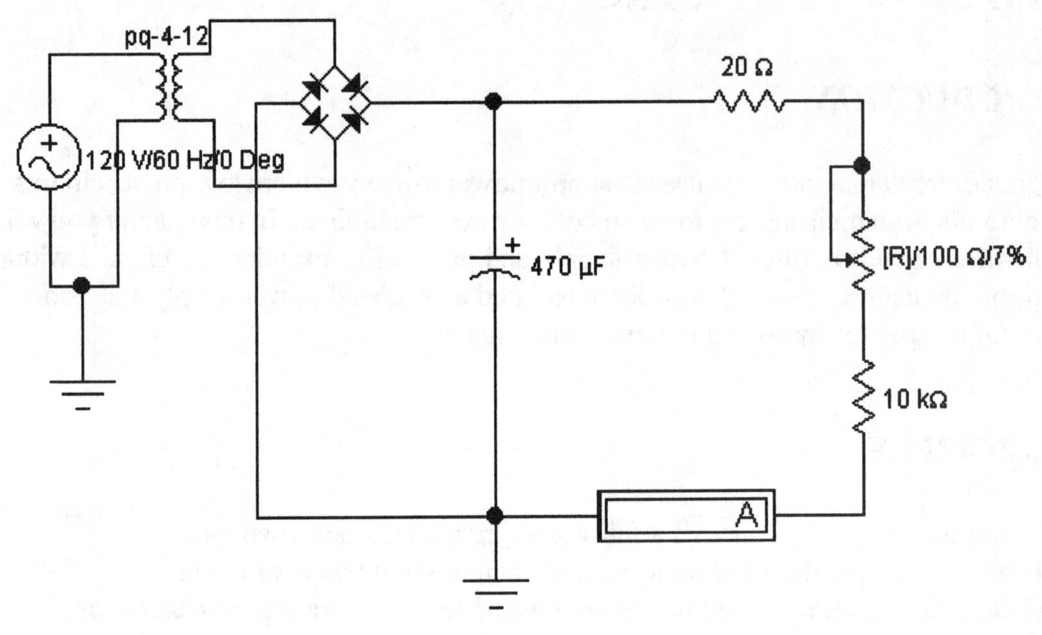

Figure 8-1

Step 2. The voltage measured across the filter capacitor will be considered the no-load voltage. Remove the 10 K-ohm resistor. Adjust the current until the ammeter reads 0.5 ampere. Record the voltage across the load (20 ohms including R). This voltage will be considered the full load voltage of this circuit. Use the oscilloscope to measure the peak-to-peak ripple voltage that is across the load. Record and show the waveform on the oscilloscope face below, Figure 8-2.

V(FL) = _______ v (p-p)= _______

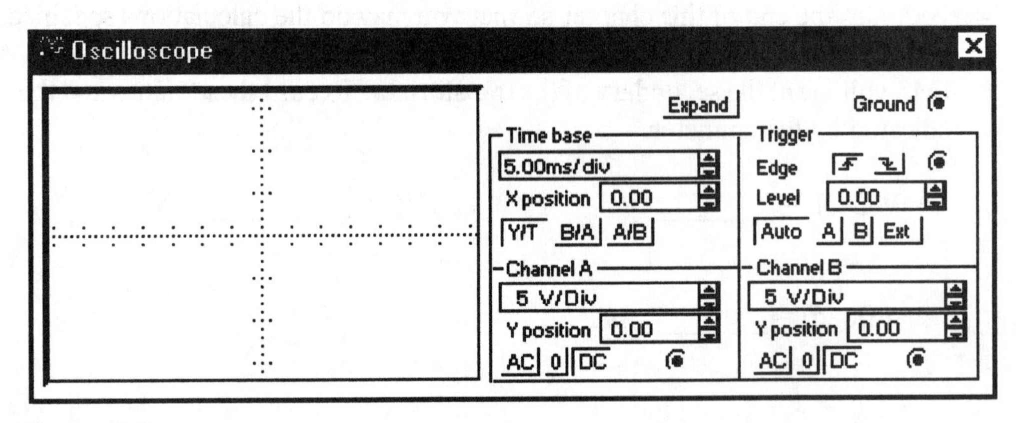

Figure 8-2

Step 3. From the formulas in the **Key Concepts** at the end of this chapter, find the output resistance of the power supply. Record.

R (out) = _______

Step 4. Adjust the load current to 250 mA. Measure the voltage across the load. Measure the peak-to-peak ripple voltage and record. Calculate the RMS value of the ripple voltage.

$V_{(DC)}$ = _______ $V_{(p\text{-}p)}$ = _______

calculated V_{rms} = _______

Step 5. From your measurements in the previous steps determine the following:

a. the load regulation % from no load to full load _______________

b. the % ripple in Step 2 _______________

c. the % ripple in Step 4 _______________

Step 6. Increase the capacitor size to 1000 uf while leaving the load the same as in Step 4. Measure the peak-to-peak ripple voltage across the load. How does the size of the ripple voltage compare to that of Step 4?

$V_{(p\text{-}p)}$ = _______

Step 7. What is the minimum voltage and maximum voltage that appear across the load in the previous step?

V_{dc} (min) = _______ V_{dc} (max) = _______

Step 8. Load the file labeled "**ex#8-1b**" from your EWB data disk. Note that the full wave rectifier circuit is the same as where you left it in the last step. Examine the circuit in Figure 8-3 and refer to the explanation of the operation in the **Key Concepts** section at the end of the chapter. Measure the DC voltage at the following points:

Vin = _______ Vz = ________ Vo = _______

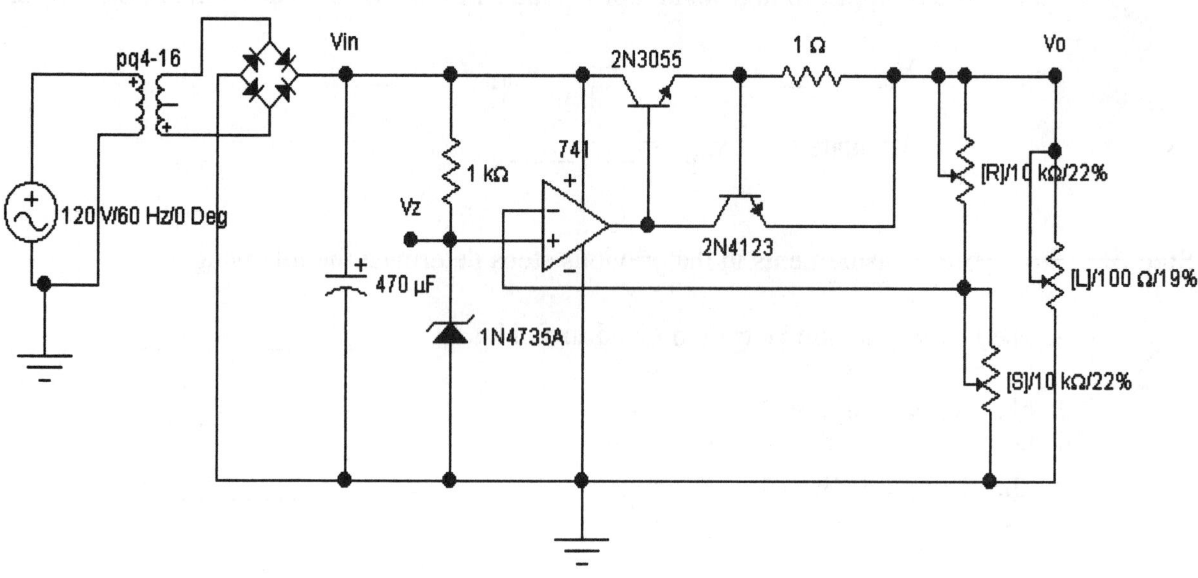

Figure 8-3

Step 9. From your measurements in Step 8, calculate the zener current, the DC current through the load, and the DC voltage at the non-inverting input of the op-amp. Record.

I z = _______ Load I = _______ N.I. V = _______

Step 10. Use the oscilloscope to measure the peak-to-peak ripple voltage at the input of the regulator and at the load. You may have difficulty measuring the output ripple, so just make an estimate.

v_{in} (p-p) = _______ v_o (p-p) = _______

Step 11. Note that the size of the current-limiting resistor is 1 ohm. Estimate the amount of current that will turn on the current-limit transistor. Place an ammeter in series with the load resistor. Slowly adjust the load resistor (key L) until the voltage at the output starts to drop. This is when the current-limiting starts to operate. Record the current and the load resistance value when this begins.

Load I = _______ Load R = _______

Step 12. Adjust the load resistor to 100-ohms maximum resistance. Change the zener to a 1N4733 (5.1-volt) or to a lower-voltage zener than is in the original circuit. Record the voltage of your zener below. Measure the new output voltage and record.

Is the output voltage twice as large as the zener voltage?

Zener Part # _______ Vz = _______

Vo = _______

Step 13. Adjust the resistance (potentiometer S) to a slightly lower resistance. What happens to the output voltage of the regulator?

Step 14. Adjust potentiometer S to a higher value than the original setting (2.2 K-ohms). What happens to the output voltage of the regulator?

Step 15. ***Hardware Circuit Application.*** Build a full-wave bridge rectifier with a regulator to meet the following specifications. You want an output voltage that will supply 9-volts to a load that requires a maximum of 500 mA. Design, build, and test, a regulated power supply using a circuit like the above simulation example. Use a commonly available 12.6 volt @ 1 ampere transformer. Build the full-wave circuit first and find the value of Ro. Choose 10% ripple for the full-wave rectifier. Record your calculations below.

Step 1 - Calculations

1. peak-to-peak ripple at the filter capacitor v_{p-p} = ________
2. capacitor size calculated, actual nearest size capacitor WVDC C = ________ ________ ________
3. Ro value from measurements Ro = ________
4. ohms value and wattage of current limit resistor Rcl = ________ @ _____ W
5. ohms value of test load resistor and wattage RL = ________ @ _____ W
6. wattage rating of series regulating transistor ____________________ W
7. wattage rating of current limit transistor ____________________ W

* Other resistors are ½ W, 5%.

Step 2. Build the circuit. *Check with your instructor before testing circuits. Observe lab safety rules.

Build the regulator circuit and test it for proper operation by using a bench power supply attached to the regulator input (without the full-wave bridge circuit). Check for the correct zener voltage to make sure it is in regulation. Check the output voltage to see if it is near your design requirement. You may use fixed resistors if potentiometers are not available.

Step 3. Measurements:

V(sec) = ________ V(cap) = ________ Vo = ________

Vz = ________ v_{in} = ________ v_o = ________

Calculate % ripple of the regulated power supply from your measurements.

% r = ________

Step 4. Load regulation test. Find the percentage load regulation by varying the load from 1 K-ohm at the output of the regulator to the maximum current of 500 mA.

Load Reg. = ________ %

Step 5. Optional test of current limiting. If you have the power resistors available, and your instructor's permission to proceed, increase the load current and determine if the current limit circuit works properly. Does the output voltage drop as you exceed your current limit? Determine whether the current limit transistor is conducting. Show your results in table form.

Exercise 8-2, Electronic Regulators

INTRODUCTION

The regulator you used in the previous exercise is packaged in a subcircuit built on your EWB data disk. This exercise has positive 12-volt, negative 12-volt, and positive 5-volt subcircuits. The subcircuits are only good when used within this exercise. You can create your own subcircuits which could be used in other circuits. Consult the EWB User Manual for further information on saving subcircuits. The subcircuit can be opened up and values adjusted to create other voltages. You will construct single- and dual-power supply circuits and will simulate these with loads attached.

Step 1. Open the circuit file labeled "**ex#8-2a**" from your EWB data disk. All you will see is a block with three terminals in the normal configuration of an electronic regulator.

Complete the circuit by adding a transformer and bridge circuit as shown in Figure 8-4.

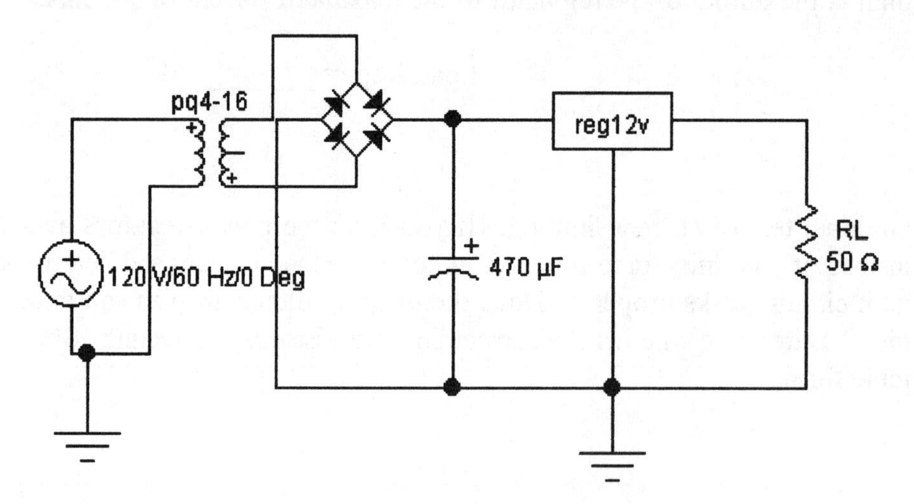

Figure 8-4

Step 2. Open up the subcircuit (double-click) and note the size of the current-limiting resistor. Close the subcircuit. Vary the load resistance RL to achieve the following currents in the table below. Measure the voltage across the filter capacitor (Vin) and the voltage across the load resistor (Vo) at each step. Plot a load regulation curve for the changes that occur at the filter capacitor.

Load current	Vin	Vo	RL
12 mA	_____	_____	_____
100 mA	_____	_____	_____
250 mA	_____	_____	_____
500 mA	_____	_____	_____

Plot the load regulation curve below.

Step 3. Construct a dual +12 V / −12 V regulator on the EWB circuit workspace from the subcircuits located in the "favorites" box on the far left of the parts bin of the file labeled "**ex#8-2b**." on your data disk, Figure 8-5.

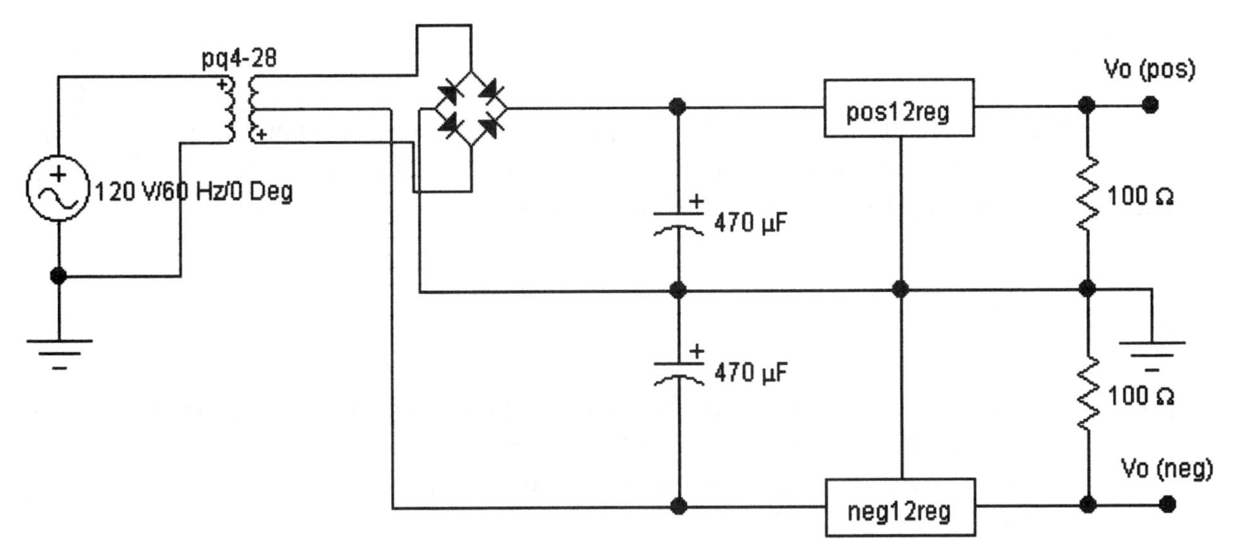

Figure 8-5

Step 4. Record the output voltages in the circuit, Figure 8-5. See if you can open up the subcircuits and change the current limits to 100 mA each. Test the revised circuit to see if both outputs limit current to 100 mA each. Do they?

Vo (+) = _______ Vo (−) = _______

Step 5. ***Optional Lab.*** Build a triple power supply with (+12 / −12 / +5) volt outputs. Use the same circuit file "**ex#8-2b**". All three regulators are in the favorites box. Show your schematic below.

Exercise 8-3
Hardware Applications Using Electronic Regulators

INTRODUCTION

Electronic regulators should have shunting capacitors at the input and output of the regulator to increase stability and ensure voltage regulation. Follow the manufacturers specifications when connecting these circuits. Construct and test the circuits below.

PROCEDURE

Step 1. Construct the 5-volt regulator as shown below, Figure 8-6. Test the regulator with a load that will carry the rated load current of 1 ampere. Measure and record the output voltage at the rated load. The regulator should be properly heat-sinked.

Vo = ________

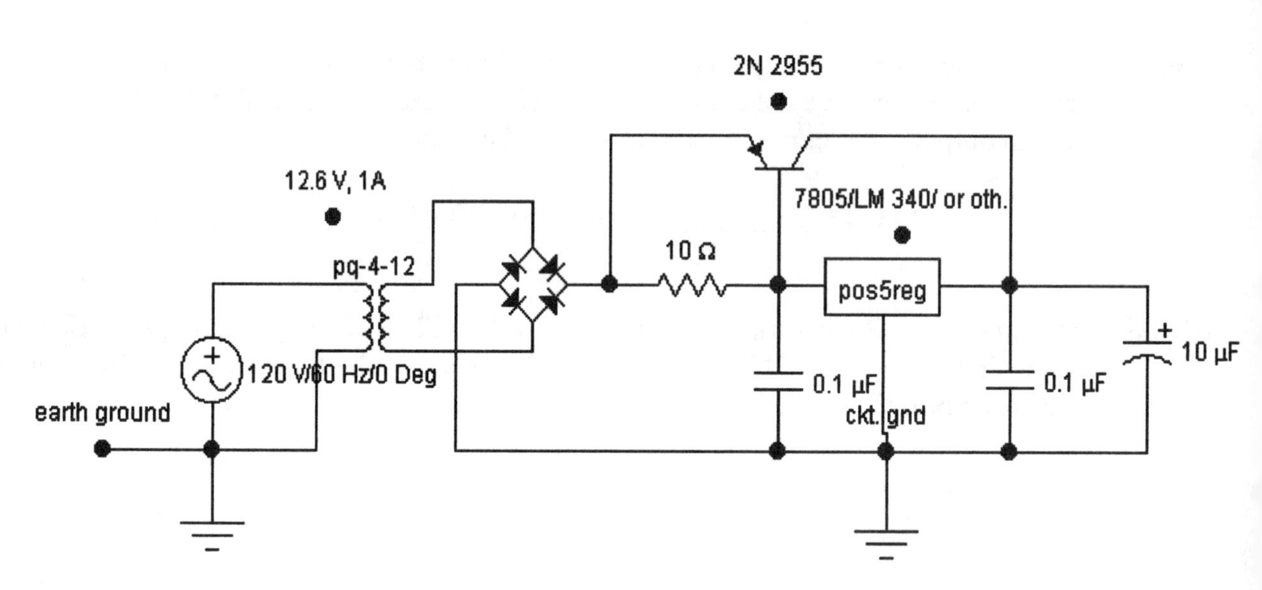

Figure 8-6

Step 2. ***Optional Circuit.*** Build and test this current boost regulator only if you have the permission of your instructor. You need to have a transformer with a current rating of 2 amperes or higher depending on what maximum current you want to supply. The 2N 3055 must have a heat sink to prevent damage. Make sure you calculate the wattage needed for the 10 ohm resistor.

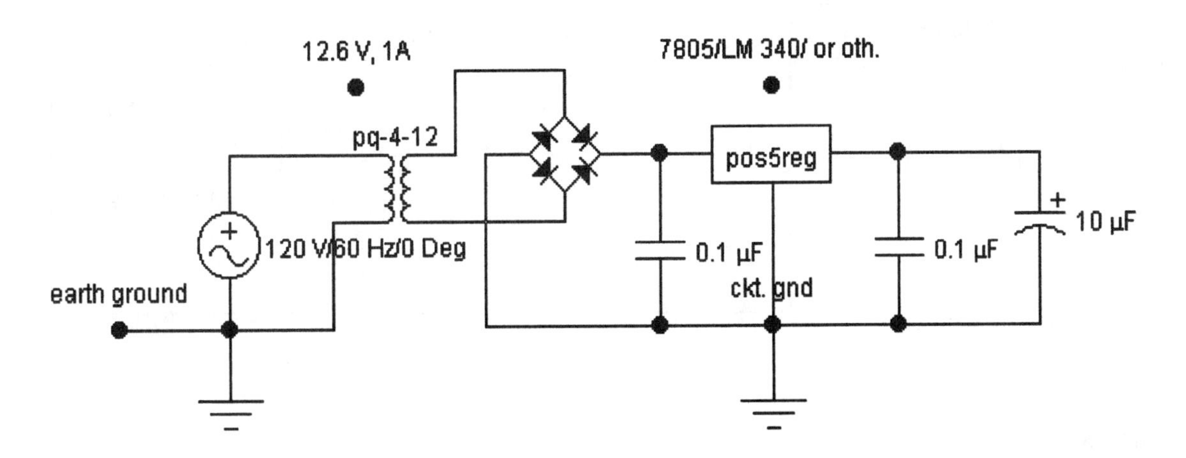

Figure 8-7

Step 3. ***Other optional circuits.*** Follow instructor guidelines for construction. Use safety precautions for laboratory work.

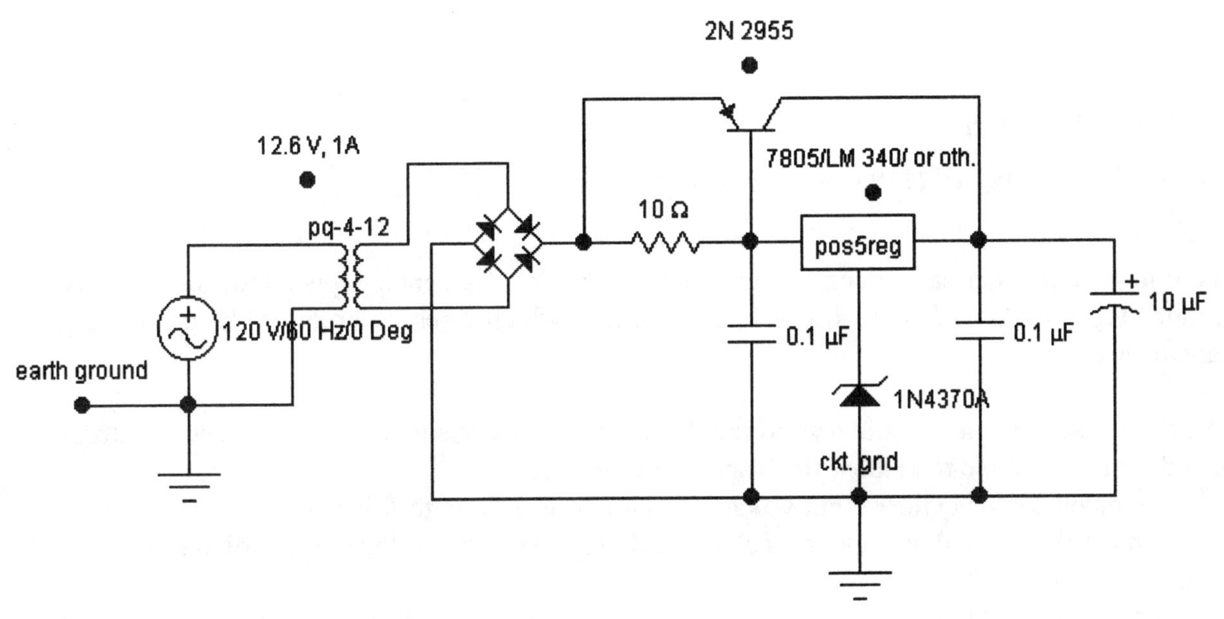

Figure 8-8

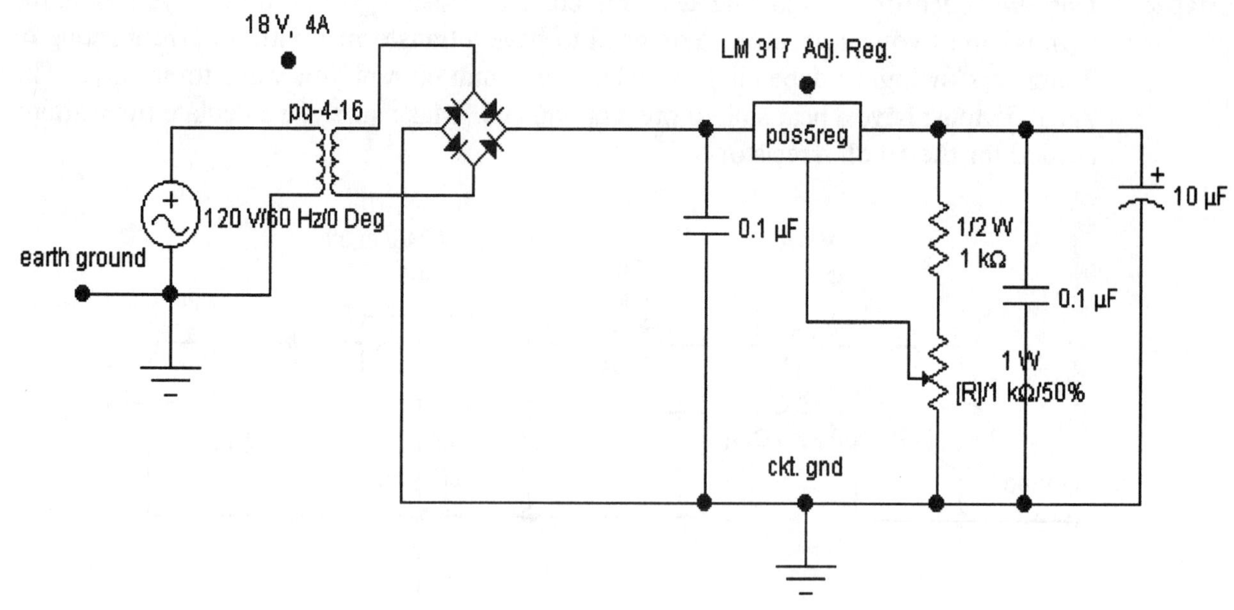

Figure 8-9

Key Concepts
Full-Wave Rectifier with Capacitor Filter

The full-wave rectifier can be constructed with two diodes if a center-tapped transformer is used. A more typical approach is to use a bridge rectifier which does not require the center-tapped transformer.

To get an exact voltage without a good regulator is not easy or usually necessary. The key items to consider in a full-wave rectifier with a capacitor filter are:

1. How much does (%) the output voltage change from no load to full load.
2. What are the rise and fall values (peak-to-peak ripple voltage) of the output voltage at the rated load current.
3. Are the ratings of the components sufficient so that the power supply does not fail under normal operating conditions? The full-wave bridge circuit and formulas are shown in Figure 8-10.

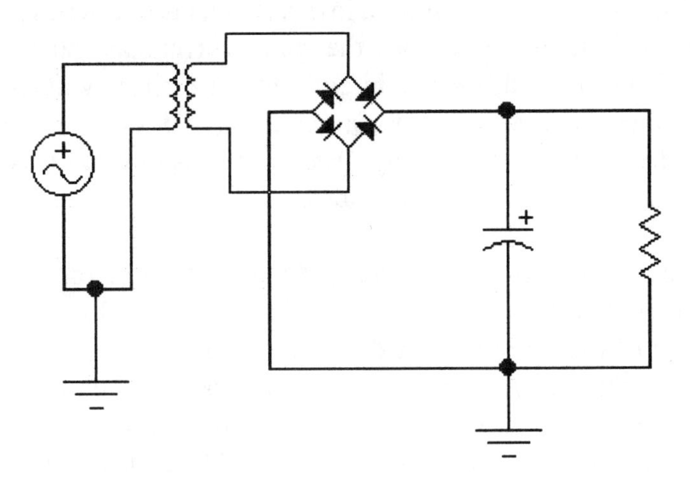

Figure 8-10

$$\% \text{ reg} = \frac{V_{NL} - V_{FL}}{V_{FL}} \times 100$$

$$V_{p-p} \text{ ripple} \cong \frac{I_L}{200C}$$

$$\%r = \frac{V_r}{V_{DC(FL)}}$$

$$V_{RMS} \text{ of ripple} = \frac{V_{p-p}}{3.5}$$

$$R_o = \frac{V_{NL} - V_{FL}}{\Delta I_L}$$

$$V_{DC(NL)} = V_{DC(FL)} + I_{L(FL)}R_o$$

Design Procedure:

1. Choose % ripple you can tolerate.
2. Solve for peak-to-peak and rms voltage of ripple.
3. Solve for capacitor size.
4. Solve for Ro either by experimentation or estimation (5-ohms to 10-ohms).
5. Solve for transformer size by solving for V(DC) no load and solving for the RMS voltage of the transformer from the formula:

$$V_{sec(RMS)} = \frac{V_{DC(NL)}}{1.4}$$

Op-Amp Regulator. The op-amp regulator used in this chapter uses a series pass transistor in series with the load resistor. This transistor carries the same current as the load resistor so power dissipation can be significant if the difference in collector-to-emitter voltage is large. The input voltage to the regulator must be at least 3-volts larger than the desired output voltage. The series pass transistor must be heat sinked accordingly. Either chassis mount or use a heat washer with silicone heat transfer compound applied to both sides.

The op-amp is used as a comparator. The zener diode voltage at one input of the op-amp is compared with the feedback voltage from a voltage divider attached to the output. Continuous comparison at the inputs of the op-amp provides base voltage control to the base of the series regulating transistor. The current-limiting transistor has its base and emitter connected across the current-limiting resistor. When the current through the limiting resistor reaches a value large enough to turn on the limiting transistor, current limiting begins. The series pass transistor sees its base emitter voltage being reduced and it begins to turn off, therefore reducing the output voltage.

When building this regulator the following steps should be followed:

1. The input voltage Vin should be 3 or 4 volts higher than the Vo desired.
2. Calculate the wattage ratings needed for all components.
3. The series pass transistor (power transistor) must be heat sinked if more than 2 watts is being dissipated.
4. The zener chosen must have a voltage rating lower than the minimum Vo desired.
5. The current-limiting resistance is determined by the formula:

$$R_{CL} = \frac{0.7\ V}{I_{L(max)}}$$

Electronic Regulators. The three-terminal regulator has an input terminal with generally the input on the left side, common in the center, and the output on the right when looking at the circuit schematic. Consult the manufacturer's specification sheet for the lead identification as this varies with the package style. Data sheets are included in Appendix A of this book.

Typical characteristics of 1A regulators are shown below.

Positive Reg.	7805	7812	7815
Negative Reg.	7905	7912	7915
Output Ro (M-ohms)	30	75	95
Load Reg. (%)	0.2	0.5	0.5
Line Reg. (%)	0.2	0.2	0.3
Ripple attenuation (dB)	70	60	60

These regulators should be wired with approximate values of capacitors as shown in Figure 8-11.

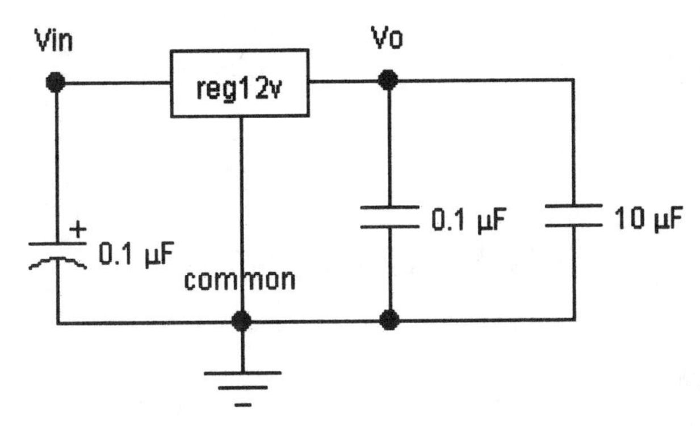

Figure 8-11

Adjustable Regulator. The LM317 adjustable regulator is designed to have a constant voltage between the output pin and the adjust pin (as are the other regulators mentioned above). The big difference is that the adjust line carries practically zero current. In the circuit in Figure 8-12, the output voltage is the sum of the voltages across R1 and R2. Since the voltage across R1 is constant at 1.2 volts, the current in this path is fixed by the value of R1. Therefore, R2 can be used to adjust the output voltage. The range of adjustment can be from 1.2 volts up to 37 volts.

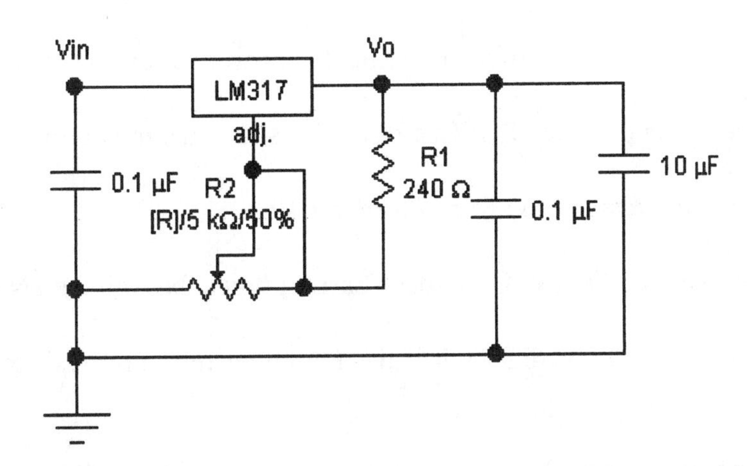

Figure 8-12

The output voltage range is determined by the following:

$$I = \frac{1.2\ V}{R1} = \frac{1.2\ V}{240} = 5\ mA \qquad V_{R2(max)} = I \times R_{2(max)} = 25\ \text{volts}$$

So:

$$V_O \cong 1.2\ V + 25\ V = 26.2\ V$$

Chapter 8 Questions
Power Supply Circuits

True or False

_____ 1. Doubling the size of a filter capacitor reduces ripple voltage by a factor of 2.

_____ 2. A transformer voltage rating is determined by the voltage it will deliver at its rated current.

_____ 3. Attaching a load resistor to a power supply will always cause a rise in the output voltage.

_____ 4. The output voltage of a well-regulated power supply is typically 100 ohms.

_____ 5. The op-amp in Step 8 of Exercise 1 is used as a comparator.

Use Figure 8-2 to answer Questions 6 through 10.

6. Load current is 500 mA. How much current flows through the 2N3055 transistor?________

7. Calculate the new current limit if the 1-ohm current limit resistor is changed to 0.5 ohms.

   ____

8. If the potentiometer (S) is made larger, the output voltage will (circle one) increase / decrease.

9. Calculate the ripple voltage at the output of the regulator if the ripple voltage at the input of the regulator is 5 v p-p. Assume the regulator has a ripple reduction of 40 dB. __________

10. Calculate the power dissipation of the series pass transistor if Vin is 18 V_{dc}, Vo is 12 V_{dc}, and the load current is 1 ampere? ____________

CHAPTER 9 Name ____________________

AUDIO AND POWER AMPLIFIERS

INTRODUCTION

Audio amplifiers consist of several stages of amplification along with tone controls, and of course, volume control. Generally, the first stage is called a preamplifier and its function consists of amplifying the voltage from one or more sources, usually more than one. Tape input, CD, microphone, or other inputs are common. Basically these signals need to have voltage amplification. The output audio circuit needs to be a power amplifier with lots of current gain, and must be a good impedance match so that the transfer of power to the speaker is efficient.

In this chapter you will examine the characteristics of each of the circuits. Furthermore, you will learn how the circuits are assembled and tested.

OBJECTIVES

- Examine the characteristics of a complementary symmetry amplifier.
- Develop a basic understanding of audio circuit troubleshooting.
- Gain an understanding of how power gain is accomplished.
- Become more familiar with calculations of db voltage, current, and power gain.
- Design, build, and test a driver circuit.
- Design, build, and test a power amplifier circuit.

Exercise 9-1, Low Power Audio Amplifier with Complementary-Symmetry Output

PROCEDURE

Step 1. Load the file labeled "**ex#9-1**" from your EWB circuits disk. This is a discrete component circuit in which you are to examine the concepts of a power amplifier. Read the **Key Concepts** explanation of this circuit at the end of this chapter. The circuit is shown below, Figure 9-1. Measure and record the following:

Q1, Q2, Q3, and Q4 (DC values)

(Q1)	VB = ________	VE = ________	VC = ________
(Q2)	VB = ________	VE = ________	VC = ________
(Q3)	VB = ________	VE = ________	VC = ________
(Q4)	VB = ________	VE = ________	VC = ________

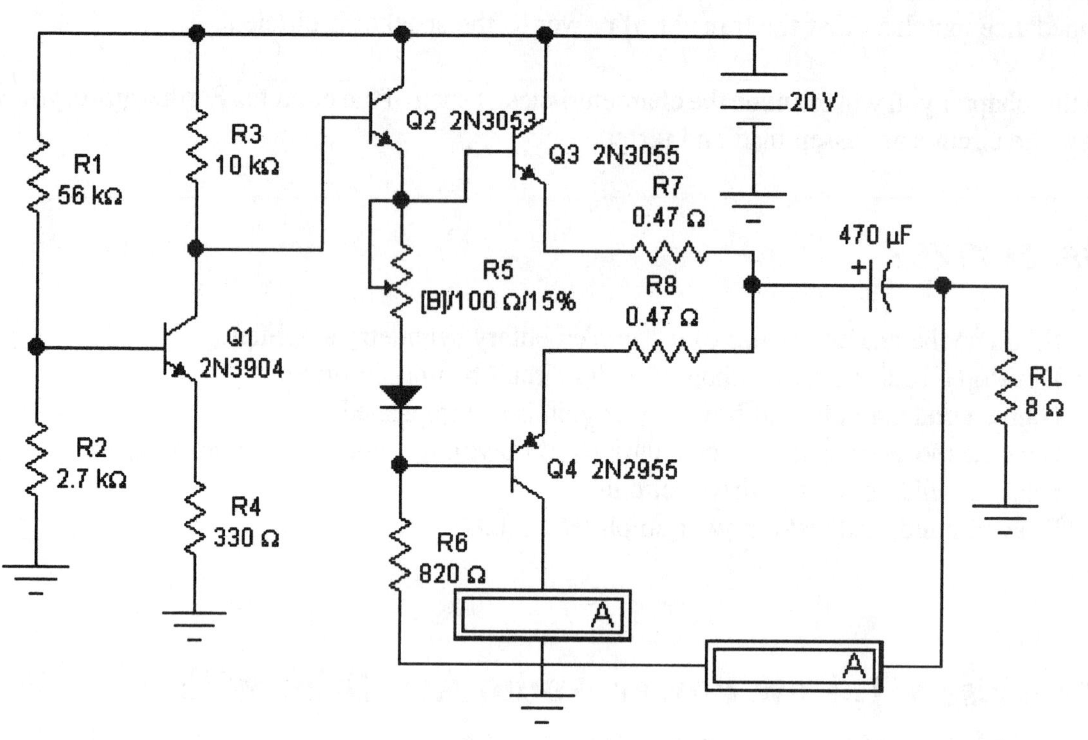

Figure 9-1

Step 2. From your measurements in Step 1, calculate the DC collector and base currents of each transistor. Then calculate the DC beta of each transistor.

	Q1	Q2	Q3	Q4
(IC)				
(IB)				
B(DC)				

Step 3. Apply a sine wave signal of 1000 mV peak-to-peak to the base of Q1 through a 10-uf coupling capacitor. Measure the peak-to-peak signal levels at the output terminal of each transistor, then determine the voltage gain of each stage. Record.

Av (Q1) = _______ Av (Q2) = _______ Av (Q3,Q4) = _______

Step 4. Increase the input signal level up to a level where distortion occurs in the output signal across the 8-ohm load. Record the peak-to-peak voltage across the load. Note the changes in load current as the amplifier is loaded down. What is the load current just prior to distortion? Calculate the maximum power (rms) developed across the load without distortion.

vo (p-p) = _______ I (load) = _______ P (load) = _______

Step 5. Change the DC biasing by changing R2 to 2.2 K-ohms. Measure the DC voltage at the junction of R7 and R8 with zero signal at the input of the amplifier. You noticed quite a change because this is a direct coupled amplifier. How does this change affect the maximum peak-to-peak output voltage. Measure and record.

V (at jct) = _______ vo (p-p) = _______

Step 6. Study the circuit and the **Key Concepts** equations and determine the input impedance and the output impedance of this complementary symmetry amplifier.

Zin = _______ Zo = _______

Step 7. Use an op-amp to replace the transistor voltage amplifier and make the other modifications as shown in Figure 9-2. The op-amp provides the voltage gain and supplies the base current needed to drive two sets of Darlington pair configurations. A single Darlington transistor can be used in low-power amplifiers. Also, notice that the large coupling capacitor is missing from the output. Measure the DC at the output terminal and record. Calculate and measure the DC bleeder current that flows up through R3, the two diodes, and R4.

Vo = _______ I (BL) = _______

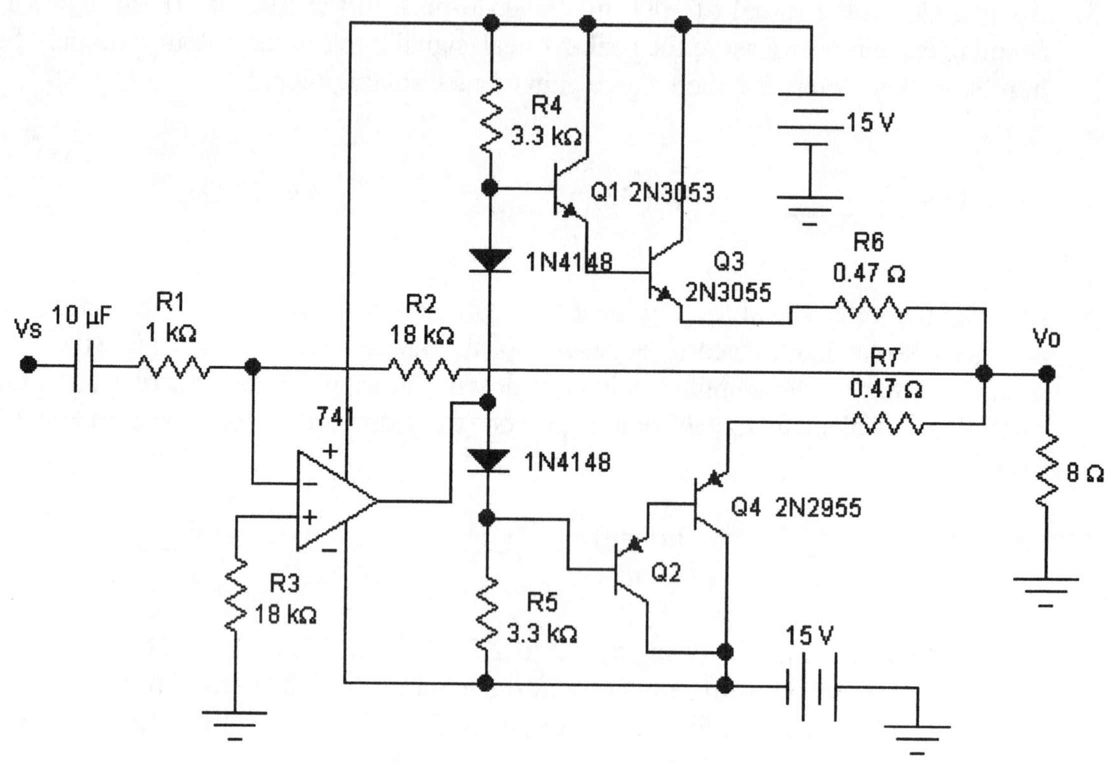

Figure 9-2

Step 8. Connect a signal generator and adjust the input signal to a 100-mV sine wave at a frequency of 1-kHz. Measure the output signal Vo and record. Calculate the voltage gain. Does the calculated voltage gain equal the measured voltage gain? Why or why not?

Vo = ________

Why/Why not?

Step 9. From your measurements in Step 8, determine the power developed across the load resistor. Increase the input signal until you see distortion in the output signal waveform. What is the maximum undistorted signal you can achieve with this circuit? Record the input signal and output signal just prior to distortion.

P (RL) = ________ Vs = ________ Vo = ________

Step 10. From your measurements above and other measurements you may choose to make, determine the following:

Number of watts the power supply would have to deliver ________ W

Number of watts Q3 and Q4 are each dissipating ________ W

Percentage efficiency (Power to RL / power supplied by source) ________ %

Exercise 9-2
Hardware Applications of Audio Amplifiers

INTRODUCTION

The LM380 is a low-power IC audio amplifier capable of delivering up to 2 W maximum into 8-ohm loads. This is a small parts count circuit than can be built easily at low cost. The IC operates best if 9-volts or more is the supply voltage and a decoupling capacitor is placed close to the power source pin on the chip. For higher power applications consult National Semiconductor Data Book or another manufacturer's literature.

If you desire to build a stereo amplifier in your lab, the LM2878 IC dual audio amplifier will deliver 5 watts/channel into 8 ohm loads. National Semiconductor's Data book for Operational Amplifiers suggests several circuits you could build. Among them are:

- a stereo amplifier with a bass tone control
- a stereo amplifier without bass control and with a servo amplifier to drive a small motor.

Step 1. Read the **Key Concepts** information and then build the circuit shown in Figure 9-3. Calculate the maximum power that can be delivered to the 8-ohm load. Remember that the power supply voltage for this circuit is 12 volts.

P (RL) = ________

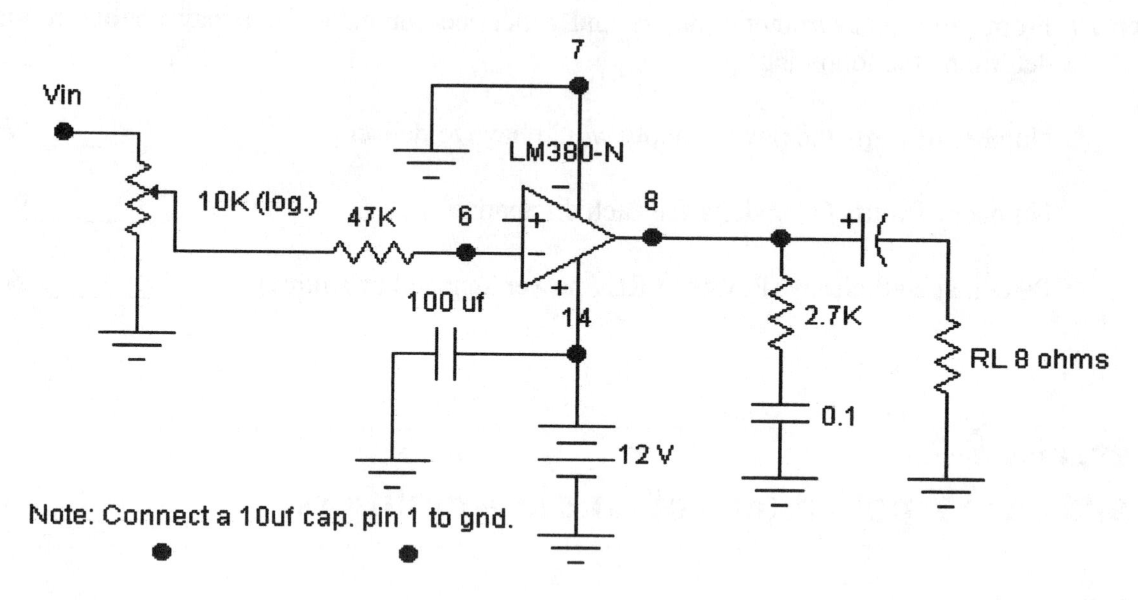

Figure 9-3

Step 2. Calculate the maximum input voltage you can apply to achieve the power across the load that you calculated in Step 1. Apply the input voltage as a 1-kHz sine wave. Measure the output voltage with an oscilloscope. What is the peak-to-peak output voltage? Calculate the power developed across the load for the measured output voltage. Does this agree with your calculation in Step 1? If not, check your method of calculation with the **Key Concepts** formula.

Vo (p-p) = _______ P (RL) = _______

Step 3. ***Optional.*** Your instructor may want you to develop a PCB layout using Electronics Workbench Layout. If you make a circuit board, make sure you connect pins 3, 4, 5, 10, 11, and 12 to the ground foil. This will help dissipate the heat so that you may achieve up to 2 watts across the speaker. An alternative is to use a clip-on heat sink that will attach to the package (24 C/W). Do all the calculations as in the previous steps, except Step 1, (calculate the peak-to-peak voltage across the load for 2 watts instead), and record below.

V (RL)p-p = _______ (calculated) V (RL)p-p = _______ (measured)

P (RL) = _______ (actual)

Part 2. Prepare a schematic, parts list, PCB layout, and written performance evaluation of this audio amplifier.

Key Concepts
Audio and Power Amplifiers

Discrete Low-Power Audio Amplifier. This amplifier can be analyzed by looking at each section and evaluating its performance. The first section is the voltage amplifier. Voltage gain in the unloaded condition is simply RC divided by RE. However, when designing this stage you need to know three things:

1. What size of load resistance will be attached to the output.
2. The input impedance to prevent serious loading of the signal.
3. The voltage gain required when the next stage is attached.

Bootstrapping the output stage increases the input impedance of this stage, and therefore reduces loading of the voltage amplifier.

Potentiometer R5 is used to turn on Q3 and Q4. It should be adjusted so that an idle current (no signal present) of about 10 mA flows. This allows the complementary output circuit to operate less AB. The idle current also reduces crossover distortion. The left side of the output capacitor should have a DC voltage of approximately one-half of the power supply voltage so that adequate peak-to-peak signal can appear across the speaker. Since this is a direct coupled amplifier, all DC voltages are set by the bias resistors R1 and R2 of the input stage. Idle current should be monitored and set prior to applying an input signal to the amplifier.

If you are building this amplifier, be sure to provide adequate heat sinks to the power transistors. The circuit and formulas to do this lab are shown in Figure 9-4 and below.

$$(Q1)\ A_V = R3 \| Z(Q2)$$

$$Z_{Q3on} = \left[B_{Q3}(R7 + X_C + RL)\right] \| (R5 + R6)$$

$$(Q2)\ Z = B\left(Z_{Q3\,on}\right)$$

$$Z_{o(amp)} = X_C + RL + \left(\frac{R3}{B_{Q2} \times B_{Q3}}\right)$$

$$(Q1)\ Z_{in(amp)} = R1 \| R2 \| B(R4 + r_e)$$

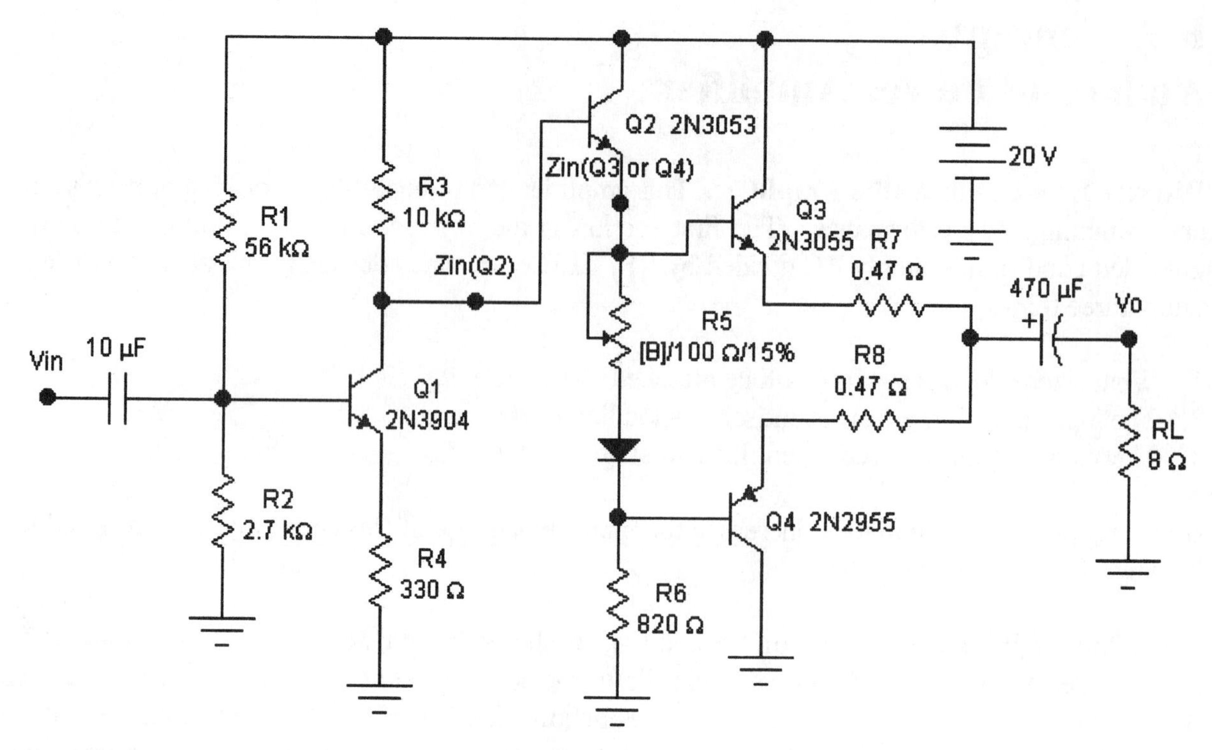

Figure 9-4

Note: See formulas below for power and efficiency calculations.

Audio Amplifier with Op-Amp. As you can see, the discrete audio amplifier requires that you design well so that loading does not affect output power. When using an op-amp as the voltage amplifier, voltage gain can be set easily and the output impedance of the op-amp is about 75-ohms. Loading, then, is rarely a problem. The circuit below simplifies the biasing with a nearly constant current circuit set up by two diodes and two resistors. The output of the op-amp is connected to the center of the two diodes, and its signal drives the Darlington pairs alternately.

The Darlington transistor pairs produce a current gain of B × B. This can amount to a typical current gain of 2500 or more. The junction of the emitter resistors and the load resistor are at approximately zero volts. This allows the signal at the output to swing from almost +15 volts to −15 volts. However, you will find that transistor and emitter resistor voltage drops will limit the peak-to-peak output voltage to much less than 30 volts peak-to-peak.

Input impedance of the amplifier is 1 K-ohm. Adjustment of Zin is easily made by changing the value of R1. Voltage gain is set by the ratio of R2 / R1. The circuit and formulas needed to do this lab are in Figure 9-5.

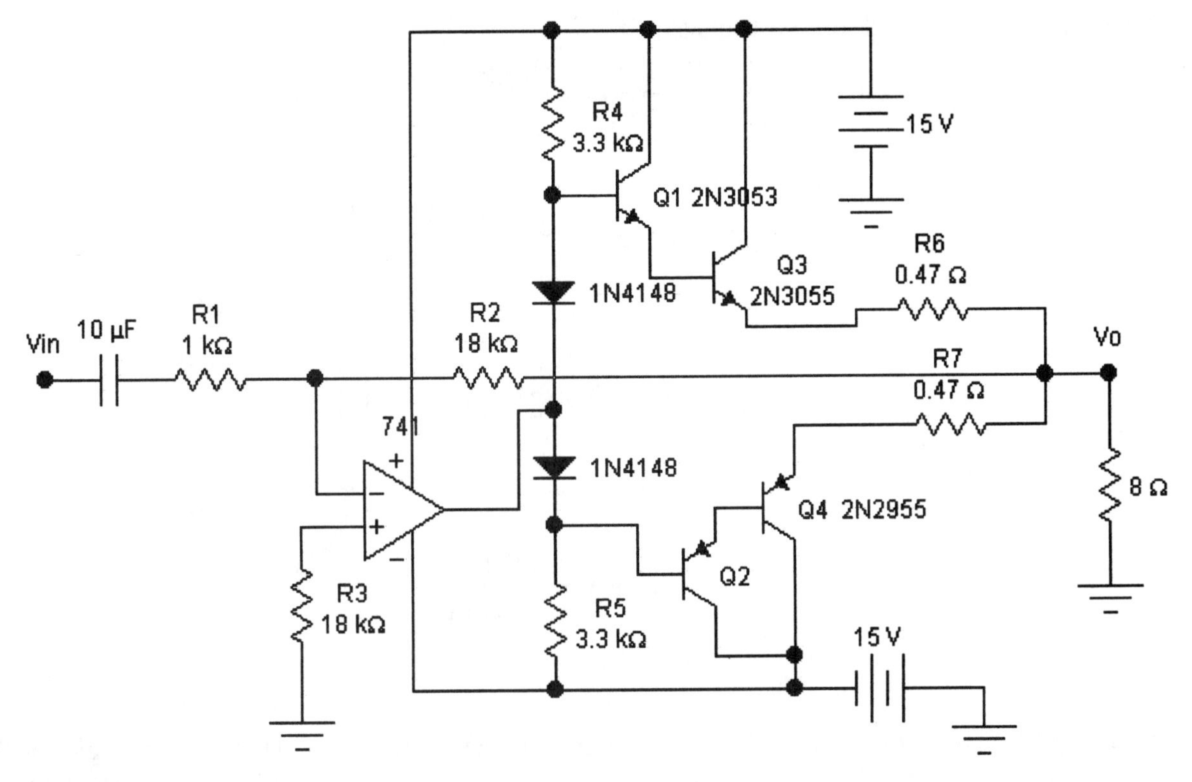

Figure 9-5

Power output to the load:

$$P_{RL} = \frac{V^2_{p-p}}{8RL} = \frac{(V_{p-p} \div 2.82)^2}{R_L}$$

Power supply dissipation:

$$P_{PSD} = V_{CC}(I_{DC} + I_{BL} + I_{idle})$$

Audio and Power Amplifiers

Power dissipated by each output transistor:

where $$I_{DC} = \frac{V_{pk}}{\pi RL}$$

$$P_{Q3} = P_{Q4} = \frac{P_{Q3+Q4} - P_{RL}}{2}$$

where

$$P_{Q3+Q4} = V_{CC}(I_{DC} + I_{idle})$$

Efficiency of output circuit

$$\% \text{ efficiency} = \frac{P_{RL}}{P_{PSD}} \times 100$$

Integrated Circuit Audio Amplifier. The integrated circuit audio amplifier comes in single and dual amplifier packages. The LM380 is one of many IC audio amplifiers ranging from 2 watts to 10's of watts. Consult a data book such as National Semiconductor's Linear IC Data book or other manufacturers of linear IC chips.

The LM380 has a fixed gain of 50 and will deliver a maximum of 5 watts to an 8-ohm load. It is recommended that a supply voltage of 9 volts or higher be used with not less than an 8-ohm load. The actual power delivered to the load depends on the load resistor and the size to the power supply voltage. The equation for power across RL is:

$$P = \frac{V^2_{CC}}{8\ RL}$$

The circuit is shown below, Figure 9-6.

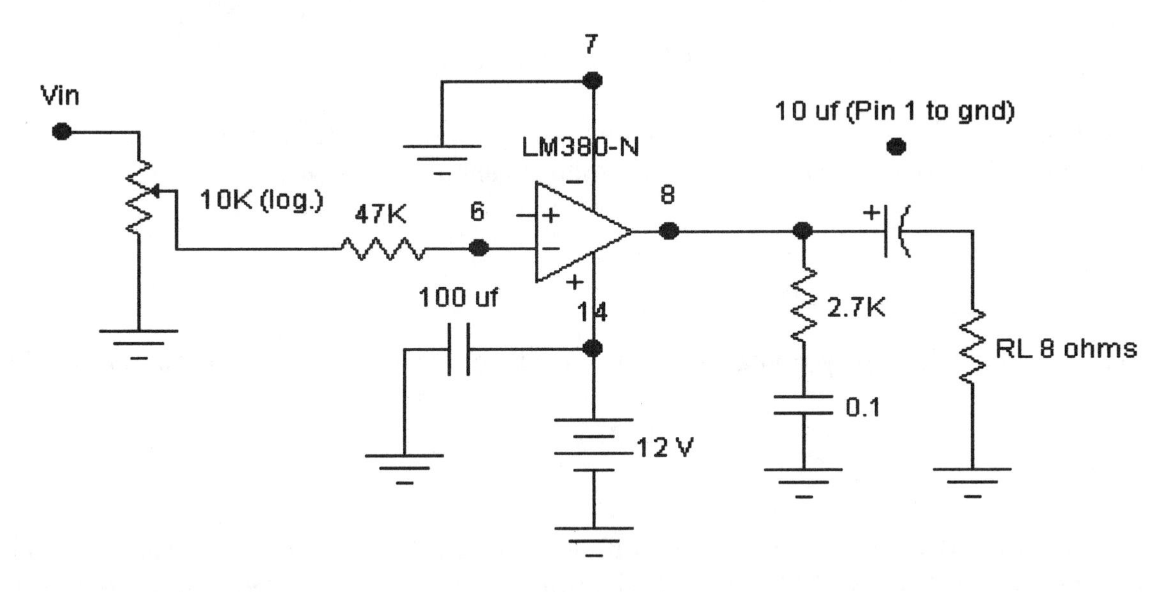

Figure 9-6

When using 12 volts or more, pins 3, 4, 5, 10, 11, and 12 should be connected to the ground PCB foil, or you should use a clip-on heat sink rated at 24 C/W.

Additional IC audio amplifier projects, including a stereo amplifier, are in Chapter 10.

Chapter 9 Questions
Audio and Power Amplifiers

True/False. Put answer on the line.

1. _____ A preamplifier stage is necessary when small signal levels are present and it is desirable to have considerable power developed at the load.

2. _____ An emitter follower stage has high input and output impedance, and therefore is needed in audio power amplifiers.

3. _____ A Darlington transistor has a current gain of approximately beta squared.

4. _____ Increasing the signal at the input of an audio amplifier does not have any effect on the power dissipation of the output transistors.

5. _____ Class A-B operation is usually 90-100% efficient.

6. A voltage amplifier, with a gain of 30, is followed by two emitter follower circuits, each with a voltage gain of 0.85. What is the overall voltage gain of this circuit?

Av ________

7. In Figure 9-3, if the supply voltage is 20 volts, the theoretical maximum power output will be

P(RL) = ________

8. An amplifier with an input impedance of 1 K-ohm is being driven by an input transducer with an internal impedance of 500 ohms. If the transducer has an unloaded output voltage of 200 mV, what voltage will the input circuit of the amplifier actually receive?

Vin = ________

9. A power amplifier has a decibel current gain of 60. Express this as a ratio of output current to input current.

____

10. What is the purpose of R5 in Figure 9-1?

CHAPTER 10
SELECTED IC PROJECTS

Name ______________________

INTRODUCTION

The final chapter of this book consists of a variety of integrated circuit projects from which you can choose depending on interest and time constraints of the course. All of the projects have been tested and built by many of my students. I have included projects for industrial applications, instrumentation, digital/microprocessor, and audio applications. All of the projects leave room for creativity and analytical thinking by the student.

Motor Control Projects. The projects shown below require you to do a little research into your previous experiences with circuits in this book, your textbook, and possibly the electronics section of the library. The first project requires you to use a digital-to-analog converter, an op-amp, and a power output circuit.

Project 10-1, Simulated Motor Speed Control

Specifications. Create a circuit that will control the speed and direction of a series DC motor. The DC motor to be driven is a 6-volt motor that is typically used to drive a tape transport in a small tape recorder or similar device. The motor will not draw more than 400 mA from the power source. The input to the DAC can be switch-controlled, or you may use a different method of your own design to improve your grade. Do not change component values inside the subcircuits!

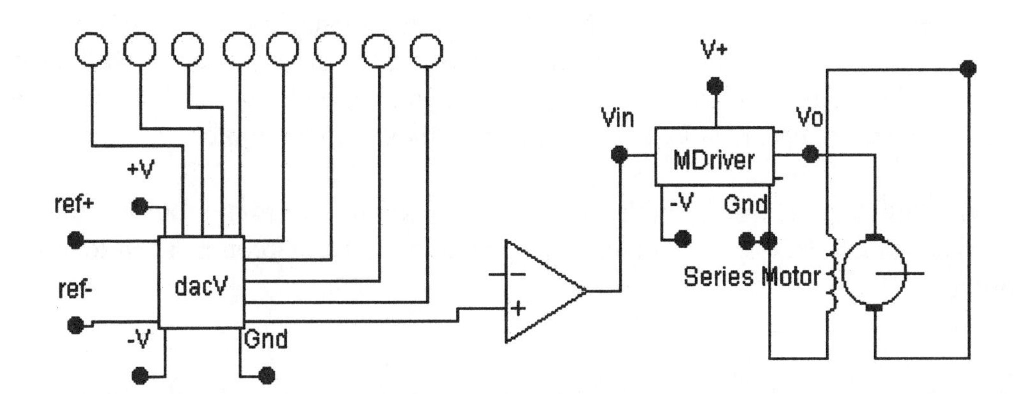

Figure 10-1

Notes on using EWB for the simulation portion of this project.

dacV block. The block labeled dacV in Figure 10-1 was created as a subcircuit in the EWB circuit file called "**ex#10-1.**" Keep in mind that this circuit can only be used with this file. The DAC is powered by ±15 volts as are the driver and the op-amp. You will need to add the power sources to the circuit. You will also need to supply a positive reference supply and a reference control rheostat. The reference voltage must be larger than the voltage required by the motor. The circles above the DAC are voltage indicator lights to show the condition of the input switches built into this device. The switches are built-in to save you wiring space. A lighted switch indicates a logic 1.

The switches are controlled from the keyboard with 7 being the most significant and 0 being the least significant bit switch respectively. The most significant bit indicator light is on the left. You may open the subcircuit and examine the internal construction. Test the DAC by itself for correct operation.

Motor Driver with Series Motor. The motor driver should look somewhat familiar to you from the power amplifier exercise. However, this is different in that we do not want both transistors conducting at the same time. The upper transistor comes on when a positive voltage appears at the input. This drives the motor in one direction at a speed which is proportional to the armature current of the motor. The armature current is determined by the voltage applied to the motor. The lower transistor drives the motor in the opposite direction when a negative voltage is applied to the input of the driver. A large DC motor should not be reversed without stopping it first. The input to the motor driver needs to be slightly larger than the 6-volts maximum delivered to the DC motor.

The motor is already connected for you in the series configuration. Consult the EWB user manual for other connections such as shunt. The motor specifications have been altered to meet the characteristics of the motor that was used in our lab. If you are building this as a hardware project you may want to alter the specifications to be similar to those of your motor.

The line protruding from the side of the armature is used to determine motor rpm. Use this connection when you are testing this circuit. Test the motor driver circuit by itself to ensure that it operates properly.

Op-Amp Circuit. This circuit acts a buffer and supplies the base current for the driver stage. This circuit should not affect the output of the DAC and should supply enough current to enable turn-on of the output transistors. Test the op-amp circuit by itself to ensure its correct operation.

Measurements.

Step 1. Check the DAC for proper operation for the following input words. Record the voltage output Vo.

0000 0000 Vo = _______ 0111 1111 Vo = _______

1000 0000 Vo = _______ 1111 1111 Vo = _______

Step 2. What voltage do you measure when checking from the armature to ground when the input word is 1111 1111? For the input word 0000 0000? Record. It should be very close to the specified voltage. If not, you need to go back and check the DAC output to see if it is working correctly.

1111 1111 V(M) = _______

0000 0000 V(M) = _______

Step 3. Make your performance tests by changing the input word while monitoring 1) voltage across the motor, 2) voltage at the rpm connector, and 3) the armature current.

Input word	V(motor)	I(motor)	V(rpm)
0000 0000			
0010 0000			
0100 0000			
0110 0000			
1000 0000			
1010 0000			
1100 0000			
1110 0000			
1111 1111			

Step 4. Plot a graph showing the change in speed of the motor versus change in armature current. Based on your graphical plot, what speed would you expect with 5-volts across this motor?

rpm = ________

Project 10-2, Hardware Applications--Speed Feedback Servo and Position Feedback Servo DC Motor Control

The following two applications use most of the components used in the previous simulation lab exercise. Build the speed feedback circuit shown in Figure 10-2. Make sure each section of the circuit works before connecting the final circuit. The tachometer generator shown below will be another 6-volt DC motor. Obtain motors that operate smoothly; for example, drive motors out of old tape players. The "tach-generator" is driven by the test motor and produces (by generator action) a feedback voltage based on speed changes of the test motor. This voltage is fed back in such a manner as to compensate for those speed variations, and it therefore will attempt to produce a nearly constant speed. An optical tachometer is needed for this experiment.

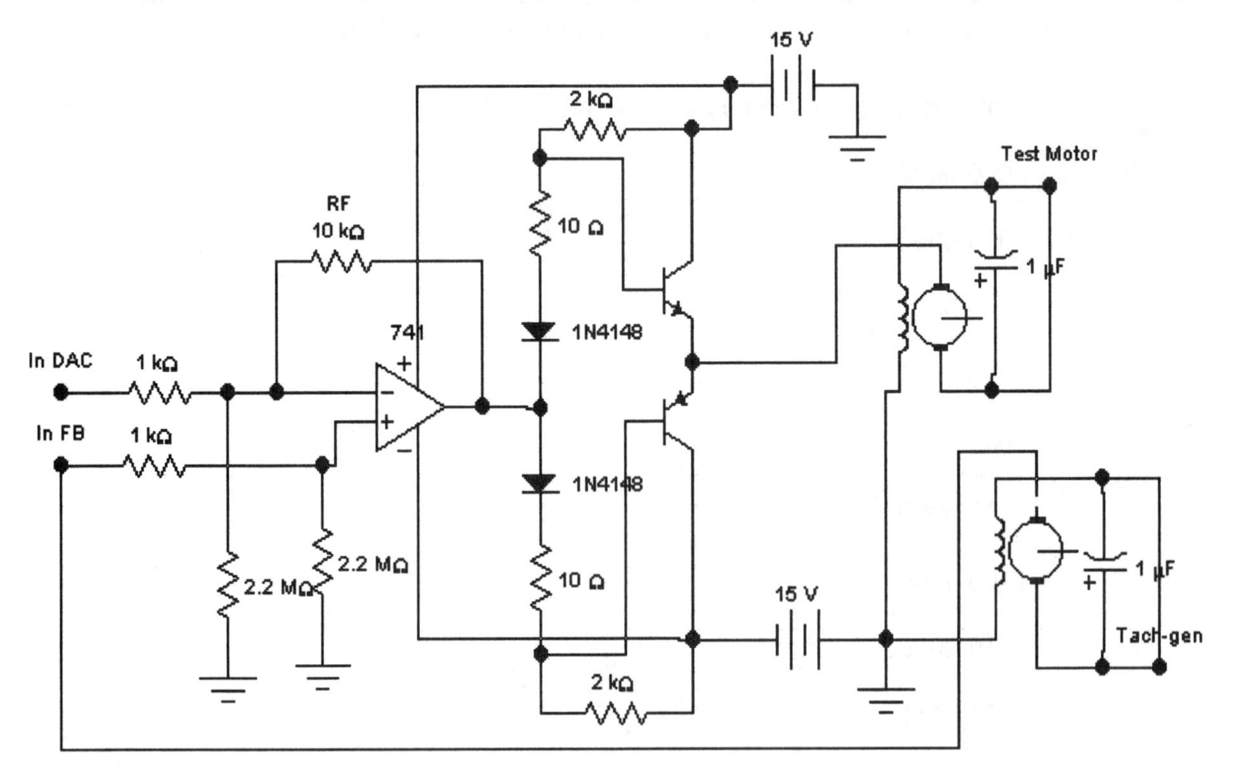

Figure 10-2

Part 1 - Speed Feedback Servo

Step 1. Connect the voltage DAC and the circuits above. Test each one individually for correct operation. The motors should be mounted on a block with the extended ends of the motor shafts facing each other. Connect the shafts together with plastic tubing (heat shrink tube or other). Power the circuit and test for smooth rotation without binding.

Step 2. The feedback signal will be a voltage generated by the driven motor. This signal is to be connected at the "FB in" point at the op-amp.

Step 3. If motor speed control cannot be obtained, switch the connection on the tachometer feedback motor. The op-amp feedback resistor RF can be adjusted for best performance.

Step 4. Adjust the DAC input digital byte to provide the first voltage specified in the table below, Chart 10-1. Record the no-load test motor drive voltage (Vmnl). Apply a small varying load to the motor shaft and record in writing how the motor drive voltage (Vm) and rpm vary with the load. Perform these steps with and without the tach-generator connected for each Vmnl in Chart 10-1.

Step 5. Record your observations in the space below the chart using the letters in the boxes of the chart as identifiers.

Step 6. Measure the tach-generator with the optical tachometer, and feedback voltage with a DMM as you vary the loading. Plot feedback voltage versus rpm on a graph. Is the tach-generator linear?

Step 7. Write down what you believe you learned from this experiment. Demonstrate this lab to your instructor.

CHART 10-1				
No load Motor	Without Feedback		With Feedback	
(Vmnl)	(open loop)		(closed loop)	
Volts	Vm vs load	rpm vs load	Vm vs load	rpm vs load
1.50	A	B	C	D
2.00	E	F	G	H
2.50	I	J	K	L
3.00	M	N	O	P

Conclusions:

Part 2, Position Feedback Servo

The position feedback servo circuit is the same as the circuit above except you do not have the feedback voltage from the tach-generator. Instead a 5 K-ohm 10-turn potentiometer is connected between the positive and negative voltage rails, with the wiper connected back to the non-inverting input resistor. Once again use the DAC-V as the input to the 1 K-ohm resistor on the inverting input of the op-amp. The circuit is in Figure 10-3 for your reference. You will adjust the feedback resistor for best response.

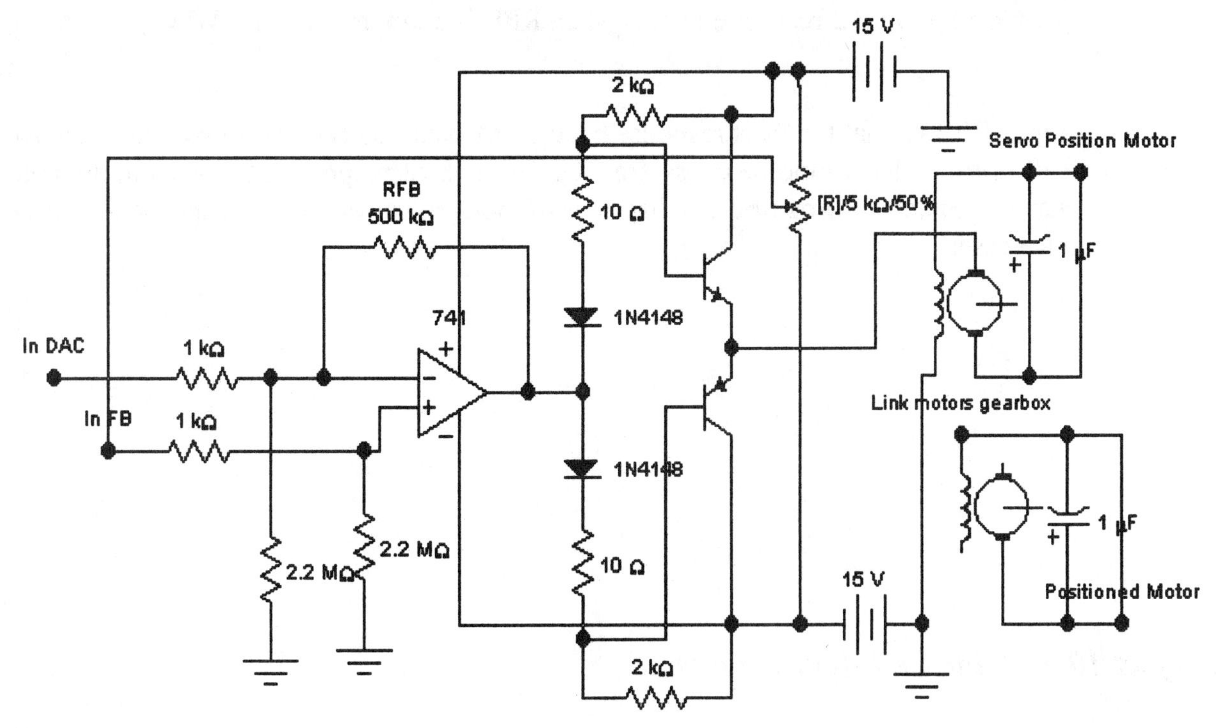

Figure 10-3

Step 1. Connect the circuit as shown in the diagram. Use a 500 K-ohm potentiometer for RFB and a 5 K-ohm, 10 turn potentiometer for fine adjustment of position. The command for position change is going to come from your voltage DAC.

Step 2. Reverse motor leads if position control cannot be attained.

Step 3. Once you can control the position, experiment with various values of RFB.

What kind of output response results when RFB is a high resistance? Why?

What kind of output response results when RFB is a low resistance? Why?

Adjust RFB so that the best response occurs. Measure RFB (power down and remove RFB first). What would be considered a good or best response of a servo to an input change? State in words or draw diagrams of input signal vs. output response for these three cases.

Project 10-3, Solar Cell Applications

The solar cell is becoming a source of electricity in remote locations, for outdoor lighting, and for recreational and consumer electronic products. Advanced manufacturing techniques have not only reduced the costs for manufacturing them but also increased their efficiency. In this project we will examine the use of a solar cell in measurement applications.

The solar cell is a very linear device when placed in the near short circuit condition. The "short circuit" condition will produce a current that is directly proportional to the amount of light energy striking the cell surface. The circuit shown below can be simulated by using a constant current source taken from the parts bin and using it as a substitute for the solar cell. The circuit is in Figure 10-4.

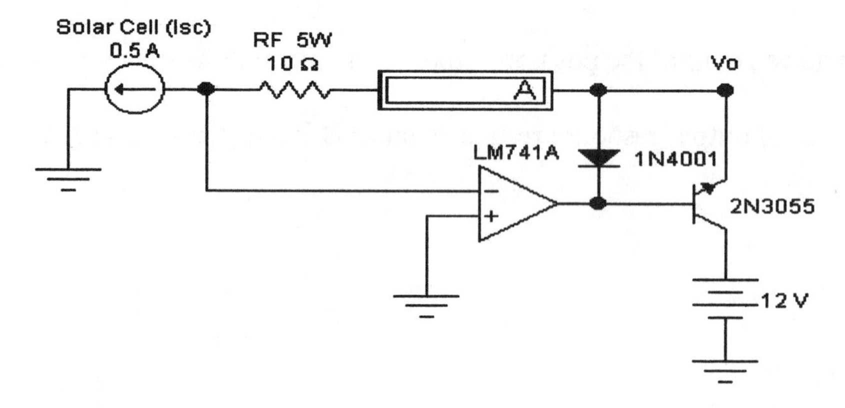

Figure 10-4

Step 1. Build and simulate the circuit above using EWB. Fill in the table below by changing the short circuit current values (Isc). Is the output voltage directly proportional to Isc?

0.05	0.1	0.15	0.2	0.25	0.3	0.35	0.4	0.45	0.5

Step 2. To test the **hardware** version of this circuit you will need to build a light source with a constant current. The lamp(s) will be a high-out infrared LED. Simulate and test the constant current circuit shown below, Figure 10-5.

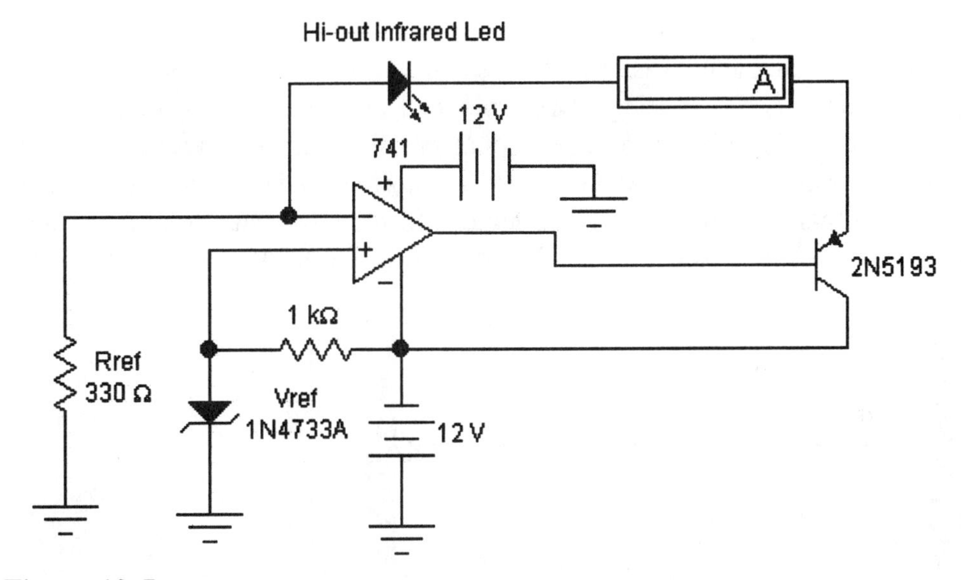

Figure 10-5

Measure and record the voltage across the zener. Vz = ________

Measure and record the voltage across Rref. Vr = ________

Calculate the current through Rref. Record. Ir = ________

Note: This is the same current as the ammeter current.

Prove that this is a constant current source by changing the resistor to 200 ohms and then measuring the current to see if it is equal to Vref divided by Rref.

Solar Cell Distance Measurement Hardware Project

This project can be built in a package of your choice. Either single-cell devices can be used or cells that have been packaged with various output current and voltage which are obviously multiple cell devices. The circuits above will work with any type of cells without major modification. In addition to the secondary vendors of components, solar cells can be purchased direct from manufacturers such as Texas Optoelectronics, Inc., and VACTEC, Inc. Addresses of these companies are in Appendix C of this book.

The high-output infrared LED is chosen because the best spectral response is in the infrared range for solar cells. A simple package for this device can be a prescription bottle or other type of pill bottle. The LED or LEDs, if using a large solar cell, and the solar cell are mounted parallel to each other at a distance of 0.25 inches or less. A plunger will act as the shade to adjust the amount of light striking the solar cell. A window should be cut into the shade which is the same shape as but just slightly smaller than the solar cell. The plunger will slide the shade from minimum exposure of the cell to light to maximum exposure of light from the LED(s). A side view is shown below.

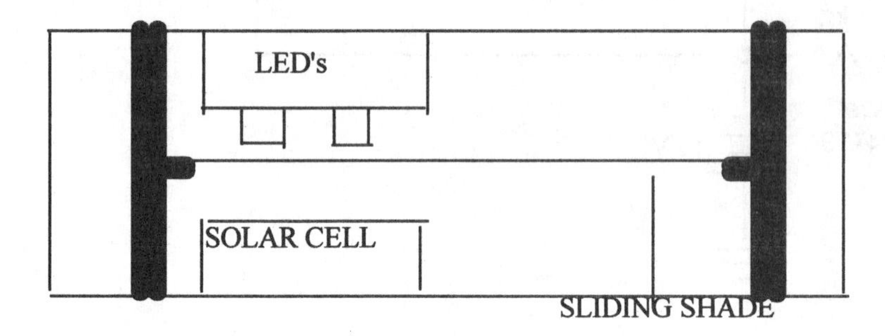

Step 1. Build the constant-current LED circuit and the solar cell circuits. Test for operation of each circuit separately and then together. Fine tuning can wait until the assembly is completed.

Step 2. Construct the package. This should be thought out carefully so it works mechanically without external light affecting the operation.

Step 3. Check the linearity of the solar cell circuit output. For example, does one-fourth movement of the shade give you one-fourth of the maximum cell voltage? Write your results in the table below.

Voltage					
Distance					

Step 4. ***Optional.*** An A/D converter can be added if desired for interface to a microcomputer or a hexadecimal display.

Project 10-4, Strain Gage Bridge Amplifier

The strain gage is a commonly used transducer used in weighing scales. The wire gage operates on the principle that if the gage is fixed on a surface, and a force of either tension (pulling) or compression (pushing together) is applied, then the resistance will increase or decrease, respectively. The change in resistance is normally only a few milliohms for a change in length of a few thousandths of an inch. A typical wire strain gage has a resistance of 120 ohms. A strain gage has a gage factor rating which is the ratio of its percent-change in resistance to its percent-change in length. Commonly used terms are strain, which is change in length divided by original length (at room temperature or other specified operating temperature), and stress, which is the amount of force acting on the unit area of the object.

The circuit shown below, Figure 10-6, is a very crude amplifier compared to more sophisticated instrumentation amplifiers and bridge circuits with multiple strain gages and temperature-sensing circuits. However, the basic principles can be simulated. A precision high-gain instrumentation amplifier made by Analog Devices, such as the AD620, can be easily used in a hardwired circuit. The gain can be controlled up to 1,000 by a single external resistor. Other manufacturers also have equally good instrumentation amplifier IC's.

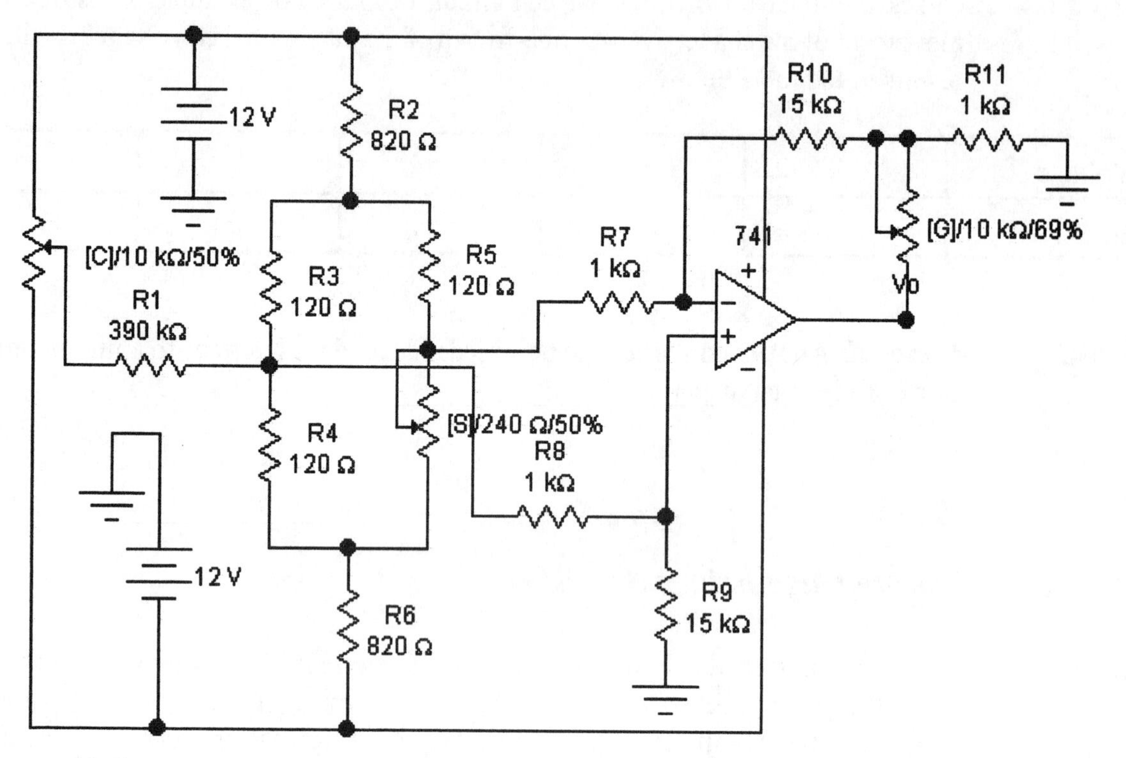

Figure 10-6

The bridge circuit is made up of three 120-ohm resistors plus a rheostat (S) which simulates a transducer. The rheostat (S) can be adjusted in 1 % increments. The gain control rheostat controls the gain of the amplifier. The gain of the amplifier shown in the schematic above is determined by the formula:

$$A_V = \frac{R_F}{R_7} \times \frac{R_{11} + R_G}{R_{11}}$$

Step 1. Calculate the voltage gain based on the values in the circuit above. Load the file labeled "**ex#10-4**" from your circuits disk. Do not change the circuit gain. Adjust the rheostat (S) to the 49% value. Assume that you want one-volt out of the op-amp for this setting (a 2.4-ohm change in resistance). Adjust the calibrate potentiometer until this is achieved. Now complete the remainder of the table.

(S) %	49 %	48 %	47 %	46 %	45 %
Vo (op-amp)					

Step 2. As you can see this would be a compression test of the specimen on which the strain gage is mounted and the values of 1 volt to 5 volts are nice values to be converted to digital values with an A/D converter. Now increase the strain gage resistance (S) from 51% to 55% (a tension force). Notice that you will have to calibrate once again to achieve a (−1) volt to a (−5) volt change. Record.

(S) %	51 %	52 %	53 %	54 %	55 %
Vo (op-amp)					

Step 3. Set the calibrate control (C) to 43% strain gage (S) to 50%. Change resistor R3 to 120.5 ohms. Record Vo. What is the voltage difference at the output terminal for the change in resistance from 120 ohms to 120.5 ohms?

Vo = _______

Change in Vo = _______

Project 10-5, Strain Gage Measurement Hardware Project

To build this project you will need a 120 ohm strain gage with leads attached. The calibration potentiometer (C) should be a 20-turn potentiometer. The LM741A can be used; however, additional gain is required to get a meaningful output. It is recommended that an instrumentation amplifier such as the AD620 be used for accuracy and more stability. Even small temperature changes will cause some instability.

This project is only meant to give you some insight into the use of the strain gage. This project is recommended by Vishay Measurements, Inc., for first-time users of strain gages. Gages can be purchased from this company and they will supply a complete kit including mounting information and adhesive. An epoxy will work if you do not wish to purchase the kit. The strain gage is mounted on an empty pop can. Prior to mounting, roughing up the surface with sandpaper is necessary. Apply the adhesive and then attach the strain gage with cellophane tape holding it in place until it is set.

Step 1. Build the circuit shown in Figure 10-7, (either the 741 or instrumentation amplifier can be used) as the amplifier.

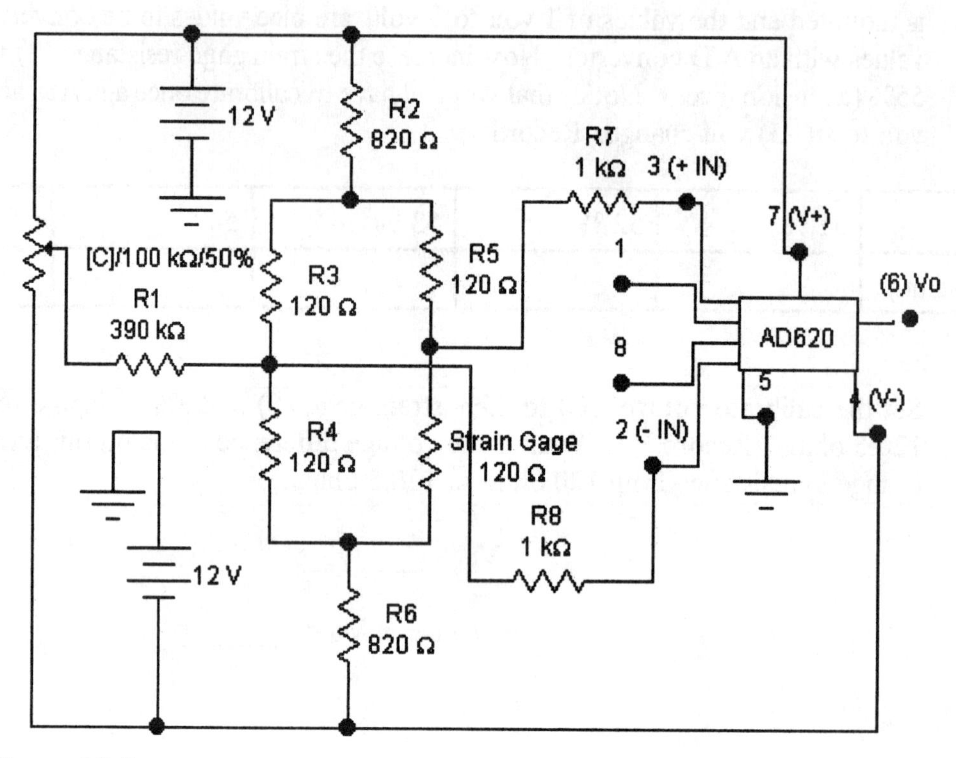

Figure 10-7

Step 2. The gain resistor is to be placed between pins 1 and 8 when using the AD620. Approximately 50 ohms will give you a gain of 1,000. Calibrate your circuit with no stress on the object (can). Measure the output voltage when the can is filled to various levels until the can is full. Your instructor will set the parameter of this laboratory depending on what components and equipment are available.

Step 3. ***Optional.*** If you are in a technical or engineering college with test equipment and a tensile testing machine, a joint project with mechanical engineering technology students using a test bar made of plastics or metal (depending on the tensile test machine available) would provide more accurate results and additional cross-discipline learning.

Other Integrated Circuit Projects (Optional)

Project 10-6, Alarm Circuits

The 555 timer is utilized in these projects. The 555 is used only to latch the LED so that it remains in the "on" condition. The operation is as follows:

1. The input signal from the sound activation of the microphone (mic) applied to the gate turns on the SCR and the LED will light after a period determined by the R and C timing components.
2. The output of the timer (pin 3) remains high, which continues to supply base current to the transistor and the lamp remains lit. This circuit is capable of operation with several types of input transducers. A simple 3-LED microphone from Radio-Shack, Digi-Key, Jameco, or other low-cost source can be used. We used an Electre[+] omnidirectional condenser microphone cartridge typical of those used in sound-controlled toys. The circuit can easily be constructed on a breadboard.

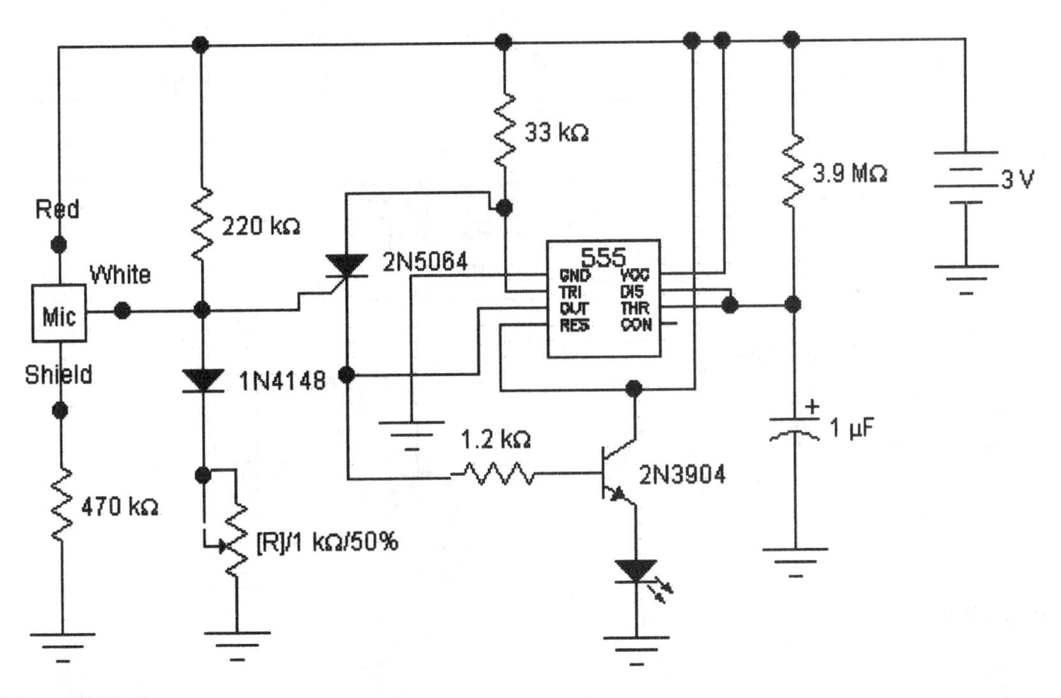

Figure 10-8

Build and test the circuit above, Figure 10-8. Experiment with the sensitivity by adjusting the rheostat (R). Try to come up with answers to the following questions:

1. Does the sound have to be aimed directly at the microphone to trigger the circuit?

2. How far away can your sound source be and still trigger the circuit? Explain your procedure and the results.

Experiment with increasing the usefulness of this circuit by using an opto-isolator or solid state relay to trigger a higher wattage lamp or a loud sounding siren or horn.

The circuit below, Figure 10-9, uses almost the same circuit as above. It is called a **touch switch.** It triggers best with a touch plate and a return wire to ground in the circuit.

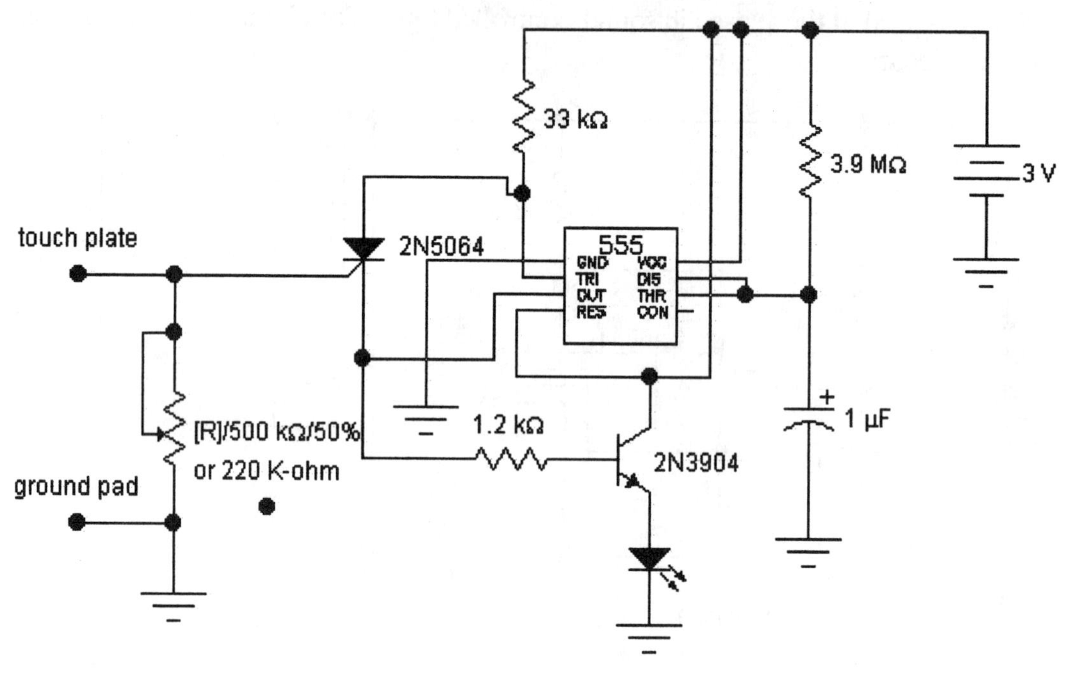

Figure 10-9

Project 10-7, Audio Amplifier Projects

The LM380 can be used to build a low-power audio amplifier. The circuits below are hardware projects that can be built quite easily with a minimum parts count.

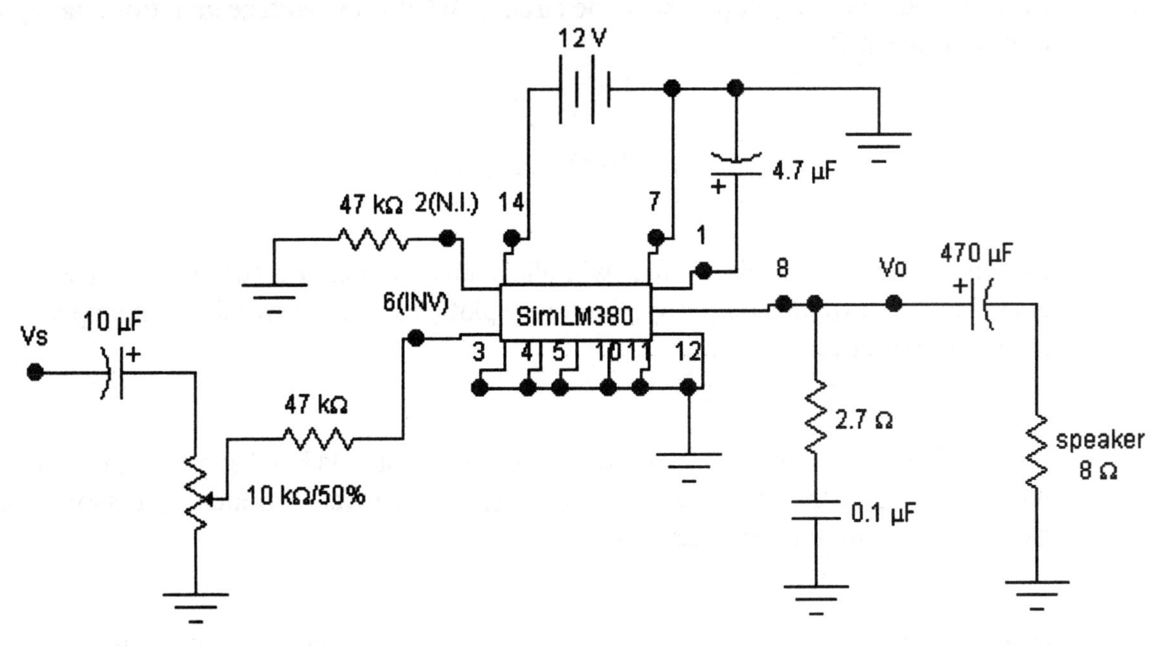

Figure 10-10

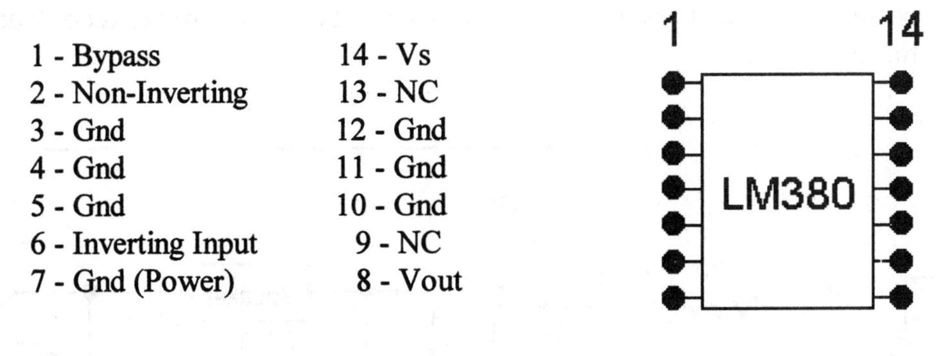

1 - Bypass	14 - Vs
2 - Non-Inverting	13 - NC
3 - Gnd	12 - Gnd
4 - Gnd	11 - Gnd
5 - Gnd	10 - Gnd
6 - Inverting Input	9 - NC
7 - Gnd (Power)	8 - Vout

Figure 10-11

Step 1. Build the hardware version of this circuit. The circuit can be breadboarded if only 9-12 volts is used. If using higher voltages, mount the components on a printed circuit board and allow at least 6 square inches of heat sink area for the pins that are grounded.

Step 2. Measure the RMS output power of the circuit. What is the voltage gain from the signal to the speaker? Record

Po = ________

Av = ________

Step 3. Design a filter for this circuit that will allow low frequencies (up to 1200 Hz) to be amplified. Do a voltage output vs. frequency plot (response curve) for your amplifier on semi-log graph paper.

Step 4. **Optional.** If time permits, a lab partner or lab group should build another amplifier that will amplify frequencies above 1200 Hz to about 5000 Hz. A frequency response plot would be done on this amplifier also.

Step 5. **Optional.** To increase power output, a bridge power amplifier may be built. A partial circuit is shown below. The maximum rated power across an 8-ohm speaker is 2.5 watts using the maximum rated supply voltage. Remember that heat sink requirements must be observed for this chip. Test the bridge circuit to see if you can achieve 5 watts of rms power across the speaker. Remember to have a large enough speaker or load test resistor for this circuit.

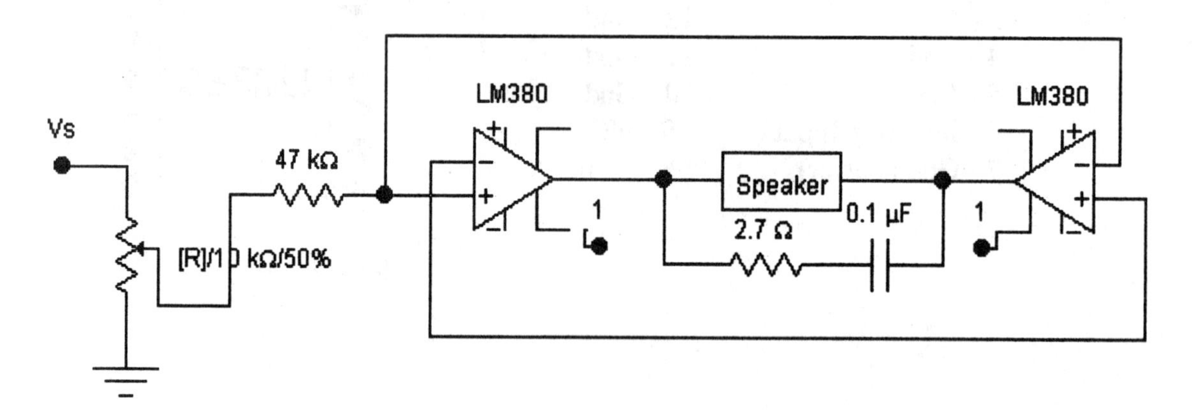

Figure 10-12

There should be zero DC current through the speaker when no signal is present. Use a test resistor before placing a speaker in this circuit. Pin 1 on each op-amp can be used to set this current to zero.

APPENDIX A
Data Sheets

National Semiconductor

ADC0801/ADC0802/ADC0803/ADC0804/ADC0805 8-Bit µP Compatible A/D Converters

General Description

The ADC0801, ADC0802, ADC0803, ADC0804 and ADC0805 are CMOS 8-bit successive approximation A/D converters that use a differential potentiometric ladder—similar to the 256R products. These converters are designed to allow operation with the NSC800 and INS8080A derivative control bus with TRI-STATE® output latches directly driving the data bus. These A/Ds appear like memory locations or I/O ports to the microprocessor and no interfacing logic is needed.

Differential analog voltage inputs allow increasing the common-mode rejection and offsetting the analog zero input voltage value. In addition, the voltage reference input can be adjusted to allow encoding any smaller analog voltage span to the full 8 bits of resolution.

Features

- Compatible with 8080 µP derivatives—no interfacing logic needed - access time - 135 ns
- Easy interface to all microprocessors, or operates "stand alone"
- Differential analog voltage inputs
- Logic inputs and outputs meet both MOS and TTL voltage level specifications
- Works with 2.5V (LM336) voltage reference
- On-chip clock generator
- 0V to 5V analog input voltage range with single 5V supply
- No zero adjust required
- 0.3" standard width 20-pin DIP package
- 20-pin molded chip carrier or small outline package
- Operates ratiometrically or with 5 V_{DC}, 2.5 V_{DC}, or analog span adjusted voltage reference

Key Specifications

- Resolution 8 bits
- Total error ±1/4 LSB, ±1/2 LSB and ±1 LSB
- Conversion time 100 µs

Typical Applications

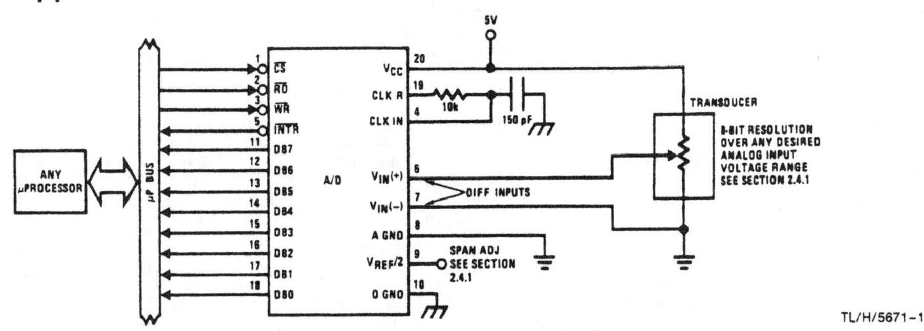

TL/H/5671-1

8080 Interface

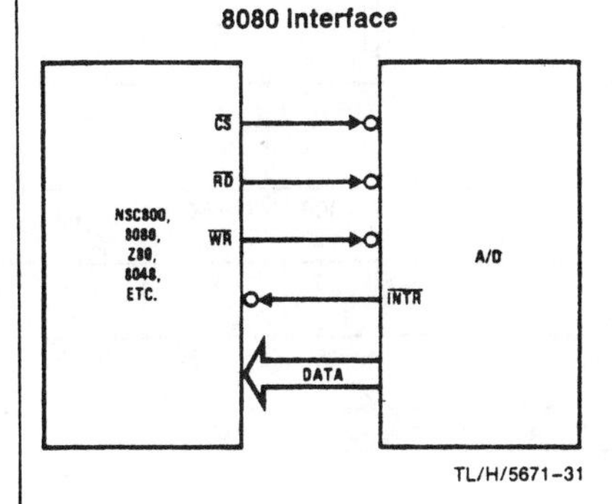

TL/H/5671-31

Error Specification (Includes Full-Scale, Zero Error, and Non-Linearity)			
Part Number	Full-Scale Adjusted	$V_{REF}/2 = 2.500\ V_{DC}$ (No Adjustments)	$V_{REF}/2$ = No Connection (No Adjustments)
ADC0801	±1/4 LSB		
ADC0802		±1/2 LSB	
ADC0803	±1/2 LSB		
ADC0804		±1 LSB	
ADC0805			±1 LSB

Absolute Maximum Ratings (Notes 1 & 2)

If Military/Aerospace specified devices are required, please contact the National Semiconductor Sales Office/Distributors for availability and specifications.

Supply Voltage (V_{CC}) (Note 3)	6.5V
Voltage	
Logic Control Inputs	−0.3V to +18V
At Other Input and Outputs	−0.3V to (V_{CC}+0.3V)
Lead Temp. (Soldering, 10 seconds)	
Dual-In-Line Package (plastic)	260°C
Dual-In-Line Package (ceramic)	300°C
Surface Mount Package	
Vapor Phase (60 seconds)	215°C
Infrared (15 seconds)	220°C
Storage Temperature Range	−65°C to +150°C
Package Dissipation at T_A=25°C	875 mW
ESD Susceptibility (Note 10)	800V

Operating Ratings (Notes 1 & 2)

Temperature Range	$T_{MIN} \le T_A \le T_{MAX}$
ADC0801/02LJ, ADC0802LJ/883	$-55°C \le T_A \le +125°C$
ADC0801/02/03/04LCJ	$-40°C \le T_A \le +85°C$
ADC0801/02/03/05LCN	$-40°C \le T_A \le +85°C$
ADC0804LCN	$0°C \le T_A \le +70°C$
ADC0802/03/04LCV	$0°C \le T_A \le +70°C$
ADC0802/03/04LCWM	$0°C \le T_A \le +70°C$
Range of V_{CC}	4.5 V_{DC} to 6.3 V_{DC}

Electrical Characteristics

The following specifications apply for V_{CC}=5 V_{DC}, $T_{MIN} \le T_A \le T_{MAX}$ and f_{CLK}=640 kHz unless otherwise specified.

Parameter	Conditions	Min	Typ	Max	Units
ADC0801: Total Adjusted Error (Note 8)	With Full-Scale Adj. (See Section 2.5.2)			±1/4	LSB
ADC0802: Total Unadjusted Error (Note 8)	V_{REF}/2=2.500 V_{DC}			±1/2	LSB
ADC0803: Total Adjusted Error (Note 8)	With Full-Scale Adj. (See Section 2.5.2)			±1/2	LSB
ADC0804: Total Unadjusted Error (Note 8)	V_{REF}/2=2.500 V_{DC}			±1	LSB
ADC0805: Total Unadjusted Error (Note 8)	V_{REF}/2-No Connection			±1	LSB
V_{REF}/2 Input Resistance (Pin 9)	ADC0801/02/03/05 ADC0804 (Note 9)	2.5 0.75	8.0 1.1		kΩ kΩ
Analog Input Voltage Range	(Note 4) V(+) or V(−)	Gnd−0.05		V_{CC}+0.05	V_{DC}
DC Common-Mode Error	Over Analog Input Voltage Range		±1/16	±1/8	LSB
Power Supply Sensitivity	V_{CC}=5 V_{DC} ±10% Over Allowed $V_{IN}(+)$ and $V_{IN}(-)$ Voltage Range (Note 4)		±1/16	±1/8	LSB

AC Electrical Characteristics

The following specifications apply for V_{CC}=5 V_{DC} and T_A=25°C unless otherwise specified.

Symbol	Parameter	Conditions	Min	Typ	Max	Units
T_C	Conversion Time	f_{CLK}=640 kHz (Note 6)	103		114	μs
T_C	Conversion Time	(Note 5, 6)	66		73	1/f_{CLK}
f_{CLK}	Clock Frequency Clock Duty Cycle	V_{CC}=5V, (Note 5) (Note 5)	100 40	640	1460 60	kHz %
CR	Conversion Rate in Free-Running Mode	$\overline{INTR}$ tied to $\overline{WR}$ with $\overline{CS}$=0 V_{DC}, f_{CLK}=640 kHz	8770		9708	conv/s
$t_{W(\overline{WR})L}$	Width of $\overline{WR}$ Input (Start Pulse Width)	$\overline{CS}$=0 V_{DC} (Note 7)	100			ns
t_{ACC}	Access Time (Delay from Falling Edge of $\overline{RD}$ to Output Data Valid)	C_L=100 pF		135	200	ns
t_{1H}, t_{0H}	TRI-STATE Control (Delay from Rising Edge of $\overline{RD}$ to Hi-Z State)	C_L=10 pF, R_L=10k (See TRI-STATE Test Circuits)		125	200	ns
t_{WI}, t_{RI}	Delay from Falling Edge of $\overline{WR}$ or $\overline{RD}$ to Reset of $\overline{INTR}$			300	450	ns
C_{IN}	Input Capacitance of Logic Control Inputs			5	7.5	pF
C_{OUT}	TRI-STATE Output Capacitance (Data Buffers)			5	7.5	pF
CONTROL INPUTS [Note: CLK IN (Pin 4) is the input of a Schmitt trigger circuit and is therefore specified separately]						
V_{IN} (1)	Logical "1" Input Voltage (Except Pin 4 CLK IN)	V_{CC}=5.25 V_{DC}	2.0		15	V_{DC}

AC Electrical Characteristics (Continued)

The following specifications apply for V_{CC} = $5V_{DC}$ and $T_{MIN} \leq T_A \leq T_{MAX}$, unless otherwise specified.

Symbol	Parameter	Conditions	Min	Typ	Max	Units
CONTROL INPUTS [Note: CLK IN (Pin 4) is the input of a Schmitt trigger circuit and is therefore specified separately]						
V_{IN} (0)	Logical "0" Input Voltage (Except Pin 4 CLK IN)	V_{CC} = 4.75 V_{DC}			0.8	V_{DC}
I_{IN} (1)	Logical "1" Input Current (All Inputs)	V_{IN} = 5 V_{DC}		0.005	1	μA_{DC}
I_{IN} (0)	Logical "0" Input Current (All Inputs)	V_{IN} = 0 V_{DC}	−1	−0.005		μA_{DC}
CLOCK IN AND CLOCK R						
V_T+	CLK IN (Pin 4) Positive Going Threshold Voltage		2.7	3.1	3.5	V_{DC}
V_T-	CLK IN (Pin 4) Negative Going Threshold Voltage		1.5	1.8	2.1	V_{DC}
V_H	CLK IN (Pin 4) Hysteresis $(V_T+)-(V_T-)$		0.6	1.3	2.0	V_{DC}
V_{OUT} (0)	Logical "0" CLK R Output Voltage	I_O = 360 μA V_{CC} = 4.75 V_{DC}			0.4	V_{DC}
V_{OUT} (1)	Logical "1" CLK R Output Voltage	I_O = −360 μA V_{CC} = 4.75 V_{DC}	2.4			V_{DC}
DATA OUTPUTS AND $\overline{INTR}$						
V_{OUT} (0)	Logical "0" Output Voltage Data Outputs $\overline{INTR}$ Output	 I_{OUT} = 1.6 mA, V_{CC} = 4.75 V_{DC} I_{OUT} = 1.0 mA, V_{CC} = 4.75 V_{DC}			 0.4 0.4	 V_{DC} V_{DC}
V_{OUT} (1)	Logical "1" Output Voltage	I_O = −360 μA, V_{CC} = 4.75 V_{DC}	2.4			V_{DC}
V_{OUT} (1)	Logical "1" Output Voltage	I_O = −10 μA, V_{CC} = 4.75 V_{DC}	4.5			V_{DC}
I_{OUT}	TRI-STATE Disabled Output Leakage (All Data Buffers)	V_{OUT} = 0 V_{DC} V_{OUT} = 5 V_{DC}	−3 		 3	μA_{DC} μA_{DC}
I_{SOURCE}		V_{OUT} Short to Gnd, T_A = 25°C	4.5	6		mA_{DC}
I_{SINK}		V_{OUT} Short to V_{CC}, T_A = 25°C	9.0	16		mA_{DC}
POWER SUPPLY						
I_{CC}	Supply Current (Includes Ladder Current) ADC0801/02/03/04LCJ/05 ADC0804LCN/LCV/LCWM	f_{CLK} = 640 kHz, $V_{REF}/2$ = NC, T_A = 25°C and $\overline{CS}$ = 5V		 1.1 1.9	 1.8 2.5	 mA mA

Note 1: Absolute Maximum Ratings indicate limits beyond which damage to the device may occur. DC and AC electrical specifications do not apply when operating the device beyond its specified operating conditions.

Note 2: All voltages are measured with respect to Gnd, unless otherwise specified. The separate A Gnd point should always be wired to the D Gnd.

Note 3: A zener diode exists, internally, from V_{CC} to Gnd and has a typical breakdown voltage of 7 V_{DC}.

Note 4: For $V_{IN}(-) \geq V_{IN}(+)$ the digital output code will be 0000 0000. Two on-chip diodes are tied to each analog input (see block diagram) which will forward conduct for analog input voltages one diode drop below ground or one diode drop greater than the V_{CC} supply. Be careful, during testing at low V_{CC} levels (4.5V), as high level analog inputs (5V) can cause this input diode to conduct–especially at elevated temperatures, and cause errors for analog inputs near full-scale. The spec allows 50 mV forward bias of either diode. This means that as long as the analog V_{IN} does not exceed the supply voltage by more than 50 mV, the output code will be correct. To achieve an absolute 0 V_{DC} to 5 V_{DC} input voltage range will therefore require a minimum supply voltage of 4.950 V_{DC} over temperature variations, initial tolerance and loading.

Note 5: Accuracy is guaranteed at f_{CLK} = 640 kHz. At higher clock frequencies accuracy can degrade. For lower clock frequencies, the duty cycle limits can be extended so long as the minimum clock high time interval or minimum clock low time interval is no less than 275 ns.

Note 6: With an asynchronous start pulse, up to 8 clock periods may be required before the internal clock phases are proper to start the conversion process. The start request is internally latched, see *Figure 2* and section 2.0.

Note 7: The $\overline{CS}$ input is assumed to bracket the $\overline{WR}$ strobe input and therefore timing is dependent on the $\overline{WR}$ pulse width. An arbitrarily wide pulse width will hold the converter in a reset mode and the start of conversion is initiated by the low to high transition of the $\overline{WR}$ pulse (see timing diagrams).

Note 8: None of these A/Ds requires a zero adjust (see section 2.5.1). To obtain zero code at other analog input voltages see section 2.5 and *Figure 5*.

Note 9: The $V_{REF}/2$ pin is the center point of a two-resistor divider connected from V_{CC} to ground. In all versions of the ADC0801, ADC0802, ADC0803, and ADC0805, and in the ADC0804LCJ, each resistor is typically 16 kΩ. In all versions of the ADC0804 except the ADC0804LCJ, each resistor is typically 2.2 kΩ.

Note 10: Human body model, 100 pF discharged through a 1.5 kΩ resistor.

National Semiconductor

DAC0800/DAC0801/DAC0802 8-Bit Digital-to-Analog Converters

General Description

The DAC0800 series are monolithic 8-bit high-speed current-output digital-to-analog converters (DAC) featuring typical settling times of 100 ns. When used as a multiplying DAC, monotonic performance over a 40 to 1 reference current range is possible. The DAC0800 series also features high compliance complementary current outputs to allow differential output voltages of 20 Vp-p with simple resistor loads as shown in *Figure 1*. The reference-to-full-scale current matching of better than ±1 LSB eliminates the need for full-scale trims in most applications while the nonlinearities of better than ±0.1% over temperature minimizes system error accumulations.

The noise immune inputs of the DAC0800 series will accept TTL levels with the logic threshold pin, V_{LC}, grounded. Changing the V_{LC} potential will allow direct interface to other logic families. The performance and characteristics of the device are essentially unchanged over the full ±4.5V to ±18V power supply range; power dissipation is only 33 mW with ±5V supplies and is independent of the logic input states.

The DAC0800, DAC0802, DAC0800C, DAC0801C and DAC0802C are a direct replacement for the DAC-08, DAC-08A, DAC-08C, DAC-08E and DAC-08H, respectively.

Features

- Fast settling output current 100 ns
- Full scale error ±1 LSB
- Nonlinearity over temperature ±0.1%
- Full scale current drift ±10 ppm/°C
- High output compliance −10V to +18V
- Complementary current outputs
- Interface directly with TTL, CMOS, PMOS and others
- 2 quadrant wide range multiplying capability
- Wide power supply range ±4.5V to ±18V
- Low power consumption 33 mW at ±5V
- Low cost

Typical Applications

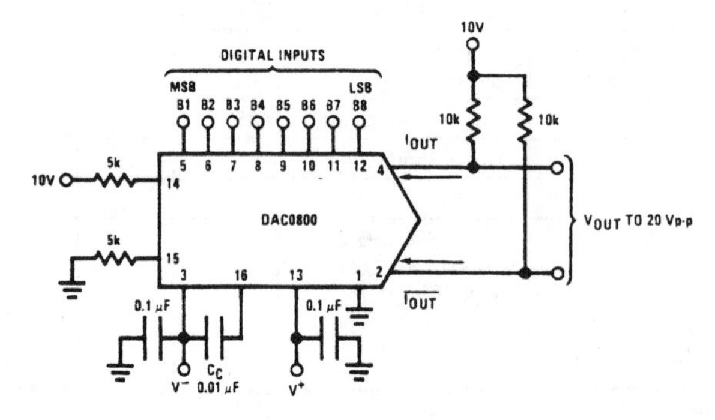

TL/H/5686-1

FIGURE 1. ±20 V_{P-P} Output Digital-to-Analog Converter (Note 4)

Ordering Information

Non-Linearity	Temperature Range	Order Numbers				
		J Package (J16A)*		N Package (N16A)*		SO Package (M16A)
±0.1% FS	0°C ≤ T_A ≤ +70°C	DAC0802LCJ	DAC-08HQ	DAC0802LCN	DAC-08HP	DAC0802LCM
±0.19% FS	−55°C ≤ T_A ≤ +125°C	DAC0800LJ	DAC-08Q			
±0.19% FS	0°C ≤ T_A ≤ +70°C	DAC0800LCJ	DAC-08EQ	DAC0800LCN	DAC-08EP	DAC0800LCM
±0.39% FS	0°C ≤ T_A ≤ +70°C			DAC0801LCN	DAC-08CP	DAC0801LCM

*Devices may be ordered by using either order number.

Absolute Maximum Ratings (Note 1)

If Military/Aerospace specified devices are required, please contact the National Semiconductor Sales Office/Distributors for availability and specifications.

Supply Voltage ($V^+ - V^-$)	±18V or 36V
Power Dissipation (Note 2)	500 mW
Reference Input Differential Voltage (V14 to V15)	V^- to V^+
Reference Input Common-Mode Range (V14, V15)	V^- to V^+
Reference Input Current	5 mA
Logic Inputs	V^- to V^- plus 36V
Analog Current Outputs ($V_S^- = -15V$)	4.25 mA
ESD Susceptibility (Note 3)	TBD V
Storage Temperature	−65°C to +150°C
Lead Temp. (Soldering, 10 seconds)	
Dual-In-Line Package (plastic)	260°C
Dual-In-Line Package (ceramic)	300°C
Surface Mount Package	
Vapor Phase (60 seconds)	215°C
Infrared (15 seconds)	220°C

Operating Conditions (Note 1)

	Min	Max	Units
Temperature (T_A)			
DAC0800L	−55	+125	°C
DAC0800LC	0	+70	°C
DAC0801LC	0	+70	°C
DAC0802LC	0	+70	°C

Electrical Characteristics

The following specifications apply for $V_S = \pm 15V$, $I_{REF} = 2$ mA and $T_{MIN} \le T_A \le T_{MAX}$ unless otherwise specified. Output characteristics refer to both I_{OUT} and $\overline{I_{OUT}}$.

Symbol	Parameter	Conditions	DAC0802LC			DAC0800L/ DAC0800LC			DAC0801LC			Units
			Min	Typ	Max	Min	Typ	Max	Min	Typ	Max	
	Resolution		8	8	8	8	8	8	8	8	8	Bits
	Monotonicity		8	8	8	8	8	8	8	8	8	Bits
	Nonlinearity				±0.1			±0.19			±0.39	%FS
t_s	Settling Time	To ±½ LSB, All Bits Switched "ON" or "OFF", T_A=25°C		100	135					100	150	ns
		DAC0800L					100	135				ns
		DAC0800LC					100	150				ns
tPLH, tPHL	Propagation Delay	T_A=25°C										
	Each Bit			35	60		35	60		35	60	ns
	All Bits Switched			35	60		35	60		35	60	ns
TCI_{FS}	Full Scale Tempco			±10	±50		±10	±50		±10	±80	ppm/°C
V_{OC}	Output Voltage Compliance	Full Scale Current Change <½ LSB, R_{OUT}>20 MΩ Typ	−10		18	−10		18	−10		18	V
I_{FS4}	Full Scale Current	V_{REF}=10.000V, R14=5.000 kΩ R15=5.000 kΩ, T_A=25°C	1.984	1.992	2.000	1.94	1.99	2.04	1.94	1.99	2.04	mA
I_{FSS}	Full Scale Symmetry	$I_{FS4}-I_{FS2}$		±0.5	±4.0		±1	±8.0		±2	±16	µA
I_{ZS}	Zero Scale Current			0.1	1.0		0.2	2.0		0.2	4.0	µA
I_{FSR}	Output Current Range	V^-=−5V	0	2.0	2.1	0	2.0	2.1	0	2.0	2.1	mA
		V^-=−8V to −18V	0	2.0	4.2	0	2.0	4.2	0	2.0	4.2	mA
	Logic Input Levels	V_{LC}=0V										
V_{IL}	Logic "0"				0.8			0.8			0.8	V
V_{IH}	Logic "1"		2.0			2.0			2.0			V
	Logic Input Current	V_{LC}=0V										
I_{IL}	Logic "0"	$-10V \le V_{IN} \le +0.8V$		−2.0	−10		−2.0	−10		−2.0	−10	µA
I_{IH}	Logic "1"	$2V \le V_{IN} \le +18V$		0.002	10		0.002	10		0.002	10	µA
V_{IS}	Logic Input Swing	V^-=−15V	−10		18	−10		18	−10		18	V
V_{THR}	Logic Threshold Range	V_S=±15V	−10		13.5	−10		13.5	−10		13.5	V
I_{15}	Reference Bias Current			−1.0	−3.0		−1.0	−3.0		−1.0	−3.0	µA
dI/dt	Reference Input Slew Rate	*(Figure 12)*	4.0	8.0		4.0	8.0		4.0	8.0		mA/µs
$PSSI_{FS+}$	Power Supply Sensitivity	$4.5V \le V+ \le 18V$		0.0001	0.01		0.0001	0.01		0.0001	0.01	%/%
$PSSI_{FS-}$		$-4.5V \le V^- \le 18V$ I_{REF}=1mA		0.0001	0.01		0.0001	0.01		0.0001	0.01	%/%
	Power Supply Current	V_S=±5V, I_{REF}=1 mA										
I+				2.3	3.8		2.3	3.8		2.3	3.8	mA
I−				−4.3	−5.8		−4.3	−5.8		−4.3	−5.8	mA
		V_S=5V, −15V, I_{REF}=2 mA										
I+				2.4	3.8		2.4	3.8		2.4	3.8	mA
I−				−6.4	−7.8		−6.4	−7.8		−6.4	−7.8	mA
		V_S=±15V, I_{REF}=2 mA										
I+				2.5	3.8		2.5	3.8		2.5	3.8	mA
I−				−6.5	−7.8		−6.5	−7.8		−6.5	−7.8	mA

Electrical Characteristics (Continued)

The following specifications apply for $V_S = \pm 15V$, $I_{REF} = 2$ mA and $T_{MIN} \le T_A \le T_{MAX}$ unless otherwise specified. Output characteristics refer to both I_{OUT} and $\overline{I_{OUT}}$.

Symbol	Parameter	Conditions	DAC0802LC			DAC0800L/ DAC0800LC			DAC0801LC			Units
			Min	Typ	Max	Min	Typ	Max	Min	Typ	Max	
P_D	Power Dissipation	$\pm 5V$, $I_{REF} = 1$ mA $5V, -15V$, $I_{REF} = 2$ mA $\pm 15V$, $I_{REF} = 2$ mA		33 108 135	48 136 174		33 108 135	48 136 174		33 108 135	48 136 174	mW mW mW

Note 1: Absolute Maximum Ratings indicate limits beyond which damage to the device may occur. DC and AC electrical specifications do not apply when operating the device beyond its specified operating conditions.

Note 2: The maximum junction temperature of the DAC0800, DAC0801 and DAC0802 is 125°C. For operating at elevated temperatures, devices in the Dual-In-Line J package must be derated based on a thermal resistance of 100°C/W, junction-to-ambient, 175°C/W for the molded Dual-In-Line N package and 100°C/W for the Small Outline M package.

Note 3: Human body model, 100 pF discharged through a 1.5 kΩ resistor.

Note 4: Pin-out numbers for the DAC080X represent the Dual-In-Line package. The Small Outline package pin-out differs from the Dual-In-Line package.

Connection Diagrams

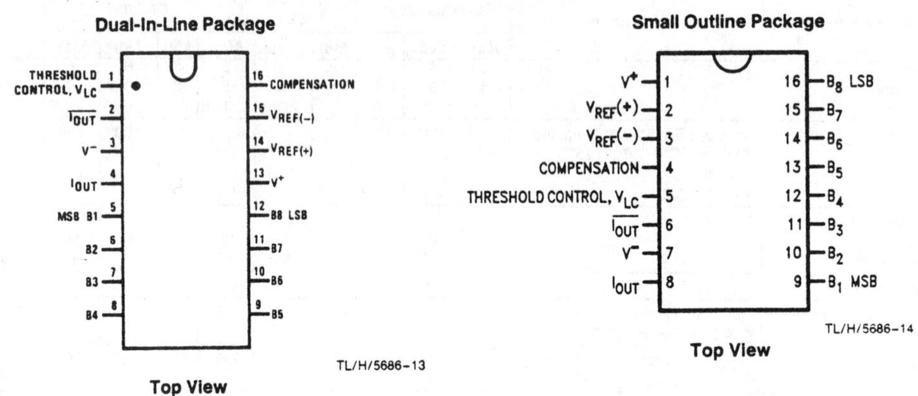

See Ordering Information

Block Diagram (Note 4)

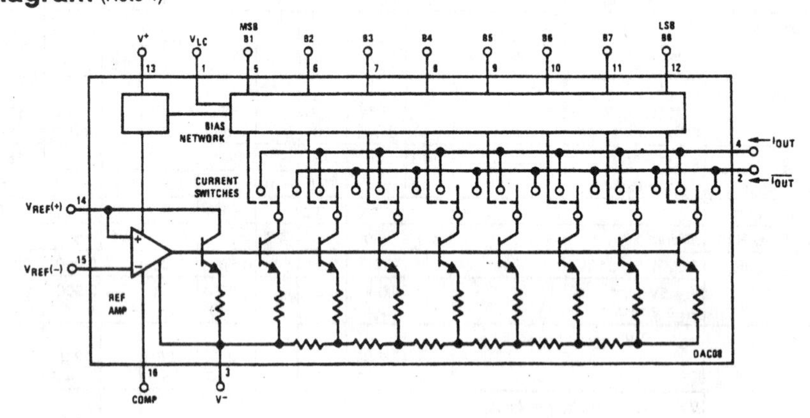

TL/H/5686-2

National Semiconductor

LM111/LM211/LM311 Voltage Comparator

General Description

The LM111, LM211 and LM311 are voltage comparators that have input currents nearly a thousand times lower than devices like the LM106 or LM710. They are also designed to operate over a wider range of supply voltages: from standard ±15V op amp supplies down to the single 5V supply used for IC logic. Their output is compatible with RTL, DTL and TTL as well as MOS circuits. Further, they can drive lamps or relays, switching voltages up to 50V at currents as high as 50 mA.

Both the inputs and the outputs of the LM111, LM211 or the LM311 can be isolated from system ground, and the output can drive loads referred to ground, the positive supply or the negative supply. Offset balancing and strobe capability are provided and outputs can be wire OR'ed. Although slower than the LM106 and LM710 (200 ns response time vs 40 ns) the devices are also much less prone to spurious oscillations. The LM111 has the same pin configuration as the LM106 and LM710.

The LM211 is identical to the LM111, except that its performance is specified over a −25°C to +85°C temperature range instead of −55°C to +125°C. The LM311 has a temperature range of 0°C to +70°C.

Features

- Operates from single 5V supply
- Input current: 150 nA max. over temperature
- Offset current: 20 nA max. over temperature
- Differential input voltage range: ±30V
- Power consumption: 135 mW at ±15V

Typical Applications**

Offset Balancing

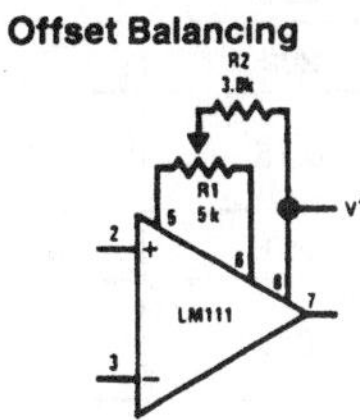

Strobing

Note: Do Not Ground Strobe Pin. Output is turned off when current is pulled from Strobe Pin.

**Note: Pin connections shown on schematic diagram and typical applications are for H08 metal can package.

Increasing Input Stage Current*

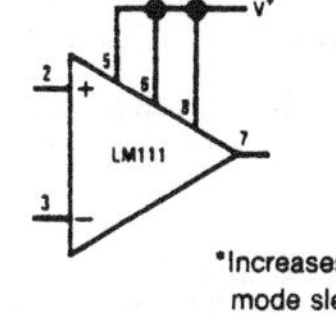

*Increases typical common mode slew from 7.0V/μs to 18V/μs.

Detector for Magnetic Transducer

Digital Transmission Isolator

Relay Driver with Strobe

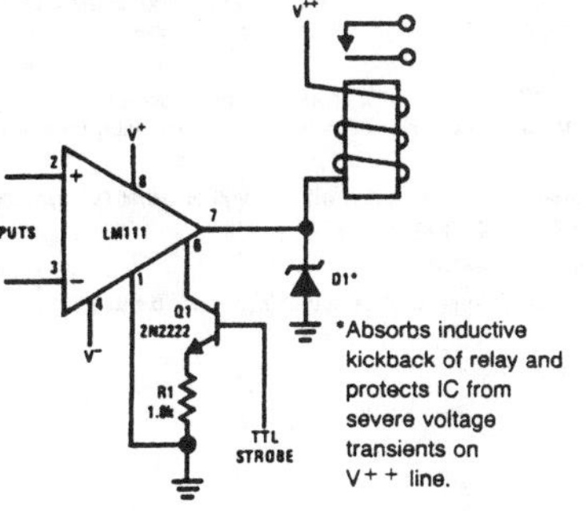

*Absorbs inductive kickback of relay and protects IC from severe voltage transients on V++ line.

Note: Do Not Ground Strobe Pin.

Strobing off Both Input* and Output Stages

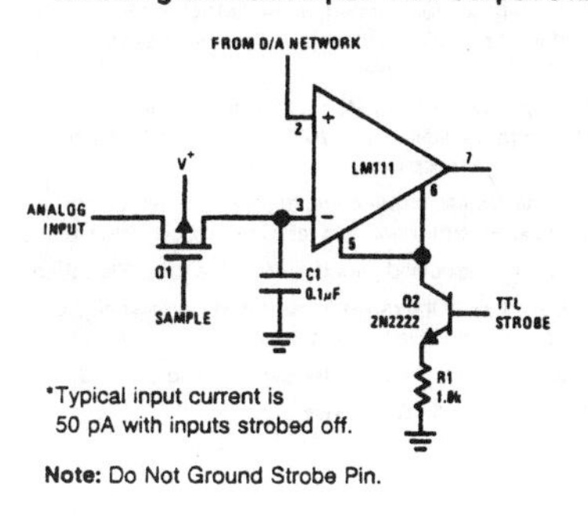

*Typical input current is 50 pA with inputs strobed off.

Note: Do Not Ground Strobe Pin.

TL/H/5704-1

Absolute Maximum Ratings for the LM111/LM211

If Military/Aerospace specified devices are required, please contact the National Semiconductor Sales Office/Distributors for availability and specifications. (Note 7)

Total Supply Voltage (V_{84}) 36V

Output to Negative Supply Voltage (V_{74}) 50V

Ground to Negative Supply Voltage (V_{14}) 30V

Differential Input Voltage ±30V

Input Voltage (Note 1) ±15V

Output Short Circuit Duration 10 sec

Operating Temperature Range LM111 −55°C to 125°C

LM211 −25°C to 85°C

Lead Temperature (Soldering, 10 sec) 260°C

Voltage at Strobe Pin $V^+ - 5V$

Soldering Information

Dual-In-Line Package

Soldering (10 seconds) ....260°C

Small Outline Package

Vapor Phase (60 seconds) ....215°C

Infrared (15 seconds) ....220°C

See AN-450 "Surface Mounting Methods and Their Effect on Product Reliability" for other methods of soldering surface mount devices.

ESD Rating (Note 8) 300V

Electrical Characteristics for the LM111 and LM211 (Note 3)

Parameter	Conditions	Min	Typ	Max	Units
Input Offset Voltage (Note 4)	$T_A = 25°C$, $R_S \leq 50k$		0.7	3.0	mV
Input Offset Current	$T_A = 25°C$		4.0	10	nA
Input Bias Current	$T_A = 25°C$		60	100	nA
Voltage Gain	$T_A = 25°C$	40	200		V/mV
Response Time (Note 5)	$T_A = 25°C$		200		ns
Saturation Voltage	$V_{IN} \leq -5$ mV, $I_{OUT} = 50$ mA $T_A = 25°C$		0.75	1.5	V
Strobe ON Current (Note 6)	$T_A = 25°C$		2.0	5.0	mA
Output Leakage Current	$V_{IN} \geq 5$ mV, $V_{OUT} = 35V$ $T_A = 25°C$, $I_{STROBE} = 3$ mA		0.2	10	nA
Input Offset Voltage (Note 4)	$R_S \leq 50$ k			4.0	mV
Input Offset Current (Note 4)				20	nA
Input Bias Current				150	nA
Input Voltage Range	$V^+ = 15V$, $V^- = -15V$, Pin 7 Pull-Up May Go To 5V	−14.5	13.8,-14.7	13.0	V
Saturation Voltage	$V^+ \geq 4.5V$, $V^- = 0$ $V_{IN} \leq -6$ mV, $I_{OUT} \leq 8$ mA		0.23	0.4	V
Output Leakage Current	$V_{IN} \geq 5$ mV, $V_{OUT} = 35V$		0.1	0.5	μA
Positive Supply Current	$T_A = 25°C$		5.1	6.0	mA
Negative Supply Current	$T_A = 25°C$		4.1	5.0	mA

Note 1: This rating applies for ±15 supplies. The positive input voltage limit is 30V above the negative supply. The negative input voltage limit is equal to the negative supply voltage or 30V below the positive supply, whichever is less.

Note 2: The maximum junction temperature of the LM111 is 150°C, while that of the LM211 is 110°C. For operating at elevated temperatures, devices in the H08 package must be derated based on a thermal resistance of 165°C/W, junction to ambient, or 20°C/W, junction to case. The thermal resistance of the dual-in-line package is 110°C/W, junction to ambient.

Note 3: These specifications apply for $V_S = \pm 15V$ and Ground pin at ground, and $-55°C \leq T_A \leq +125°C$, unless otherwise stated. With the LM211, however, all temperature specifications are limited to $-25°C \leq T_A \leq +85°C$. The offset voltage, offset current and bias current specifications apply for any supply voltage from a single 5V supply up to ±15V supplies.

Note 4: The offset voltages and offset currents given are the maximum values required to drive the output within a volt of either supply with a 1 mA load. Thus, these parameters define an error band and take into account the worst-case effects of voltage gain and R_S.

Note 5: The response time specified (see definitions) is for a 100 mV input step with 5 mV overdrive.

Note 6: This specification gives the range of current which must be drawn from the strobe pin to ensure the output is properly disabled. Do not short the strobe pin to ground; it should be current driven at 3 to 5 mA.

Note 7: Refer to RETS111X for the LM111H, LM111J and LM111J-8 military specifications.

Note 8: Human body model, 1.5 kΩ in series with 100 pF.

Absolute Maximum Ratings for the LM311

If Military/Aerospace specified devices are required, please contact the National Semiconductor Sales Office/Distributors for availability and specifications.

Total Supply Voltage (V_{84})	36V
Output to Negative Supply Voltage V_{74})	40V
Ground to Negative Supply Voltage V_{14})	30V
Differential Input Voltage	±30V
Input Voltage (Note 1)	±15V
Power Dissipation (Note 2)	500 mW
ESD Rating (Note 7)	300V
Output Short Circuit Duration	10 sec
Operating Temperature Range	0° to 70°C
Storage Temperature Range	−65°C to 150°C
Lead Temperature (soldering, 10 sec)	260°C
Voltage at Strobe Pin	$V^{+}-5V$

Soldering Information
- Dual-In-Line Package
  - Soldering (10 seconds) ....260°C
- Small Outline Package
  - Vapor Phase (60 seconds) ....215°C
  - Infrared (15 seconds) ....220°C

See AN-450 "Surface Mounting Methods and Their Effect on Product Reliability" for other methods of soldering surface mount devices.

Electrical Characteristics for the LM311 (Note 3)

Parameter	Conditions	Min	Typ	Max	Units
Input Offset Voltage (Note 4)	$T_A=25°C$, $R_S \le 50k$		2.0	7.5	mV
Input Offset Current (Note 4)	$T_A=25°C$		6.0	50	nA
Input Bias Current	$T_A=25°C$		100	250	nA
Voltage Gain	$T_A=25°C$	40	200		V/mV
Response Time (Note 5)	$T_A=25°C$		200		ns
Saturation Voltage	$V_{IN} \le -10$ mV, $I_{OUT}=50$ mA $T_A=25°C$		0.75	1.5	V
Strobe ON Current (Note 6)	$T_A=25°C$		2.0	5.0	mA
Output Leakage Current	$V_{IN} \ge 10$ mV, $V_{OUT}=35V$ $T_A=25°C$, $I_{STROBE}=3$ mA V^{-} = Pin 1 = −5V		0.2	50	nA
Input Offset Voltage (Note 4)	$R_S \le 50K$			10	mV
Input Offset Current (Note 4)				70	nA
Input Bias Current				300	nA
Input Voltage Range		−14.5	13.8, −14.7	13.0	V
Saturation Voltage	$V^{+} \ge 4.5V$, $V^{-}=0$ $V_{IN} \le -10$ mV, $I_{OUT} \le 8$ mA		0.23	0.4	V
Positive Supply Current	$T_A=25°C$		5.1	7.5	mA
Negative Supply Current	$T_A=25°C$		4.1	5.0	mA

Note 1: This rating applies for ±15V supplies. The positive input voltage limit is 30V above the negative supply. The negative input voltage limit is equal to the negative supply voltage or 30V below the positive supply, whichever is less.

Note 2: The maximum junction temperature of the LM311 is 110°C. For operating at elevated temperature, devices in the H08 package must be derated based on a thermal resistance of 165°C/W, junction to ambient, or 20°C/W, junction to case. The thermal resistance of the dual-in-line package is 100°C/W, junction to ambient.

Note 3: These specifications apply for $V_S = \pm 15V$ and Pin 1 at ground, and $0°C < T_A < +70°C$, unless otherwise specified. The offset voltage, offset current and bias current specifications apply for any supply voltage from a single 5V supply up to ±15V supplies.

Note 4: The offset voltages and offset currents given are the maximum values required to drive the output within a volt of either supply with 1 mA load. Thus, these parameters define an error band and take into account the worst-case effects of voltage gain and R_S.

Note 5: The response time specified (see definitions) is for a 100 mV input step with 5 mV overdrive.

Note 6: This specification gives the range of current which must be drawn from the strobe pin to ensure the output is properly disabled. Do not short the strobe pin to ground; it should be current driven at 3 to 5 mA.

Note 7: Human body model, 1.5 kΩ in series with 100 pF.

National Semiconductor

LM35/LM35A/LM35C/LM35CA/LM35D Precision Centigrade Temperature Sensors

General Description

The LM35 series are precision integrated-circuit temperature sensors, whose output voltage is linearly proportional to the Celsius (Centigrade) temperature. The LM35 thus has an advantage over linear temperature sensors calibrated in ° Kelvin, as the user is not required to subtract a large constant voltage from its output to obtain convenient Centigrade scaling. The LM35 does not require any external calibration or trimming to provide typical accuracies of ±¼°C at room temperature and ±¾°C over a full −55 to +150°C temperature range. Low cost is assured by trimming and calibration at the wafer level. The LM35's low output impedance, linear output, and precise inherent calibration make interfacing to readout or control circuitry especially easy. It can be used with single power supplies, or with plus and minus supplies. As it draws only 60 μA from its supply, it has very low self-heating, less than 0.1°C in still air. The LM35 is rated to operate over a −55° to +150°C temperature range, while the LM35C is rated for a −40° to +110°C range (−10° with improved accuracy). The LM35 series is available packaged in hermetic TO-46 transistor packages, while the LM35C, LM35CA, and LM35D are also available in the plastic TO-92 transistor package. The LM35D is also available in an 8-lead surface mount small outline package and a plastic TO-202 package.

Features

- Calibrated directly in ° Celsius (Centigrade)
- Linear + 10.0 mV/°C scale factor
- 0.5°C accuracy guaranteeable (at +25°C)
- Rated for full −55° to +150°C range
- Suitable for remote applications
- Low cost due to wafer-level trimming
- Operates from 4 to 30 volts
- Less than 60 μA current drain
- Low self-heating, 0.08°C in still air
- Nonlinearity only ±¼°C typical
- Low impedance output, 0.1 Ω for 1 mA load

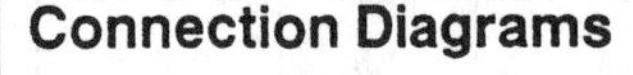

Connection Diagrams

TO-46
Metal Can Package*

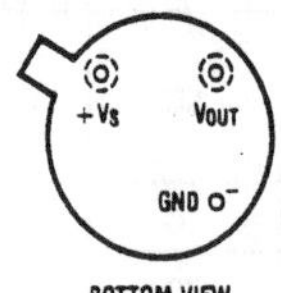

TL/H/5516-1

*Case is connected to negative pin (GND)

Order Number LM35H, LM35AH, LM35CH, LM35CAH or LM35DH
See NS Package Number H03H

TO-92
Plastic Package

TL/H/5516-2

Order Number LM35CZ, LM35CAZ or LM35DZ
See NS Package Number Z03A

SO-8
Small Outline Molded Package

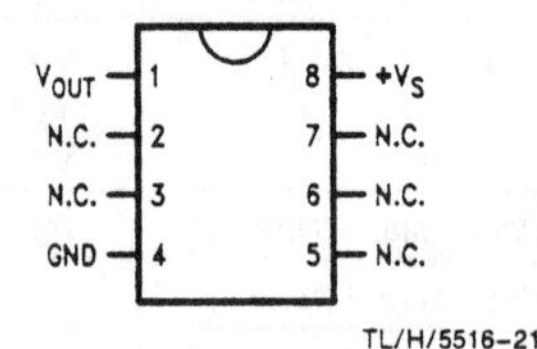

TL/H/5516-21

Top View

N.C. = No Connection

Order Number LM35DM
See NS Package Number M08A

TO-202
Plastic Package

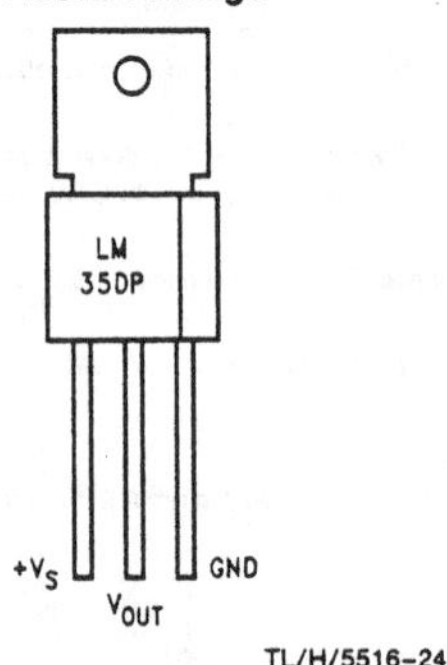

TL/H/5516-24

Order Number LM35DP
See NS Package Number P03A

Typical Applications

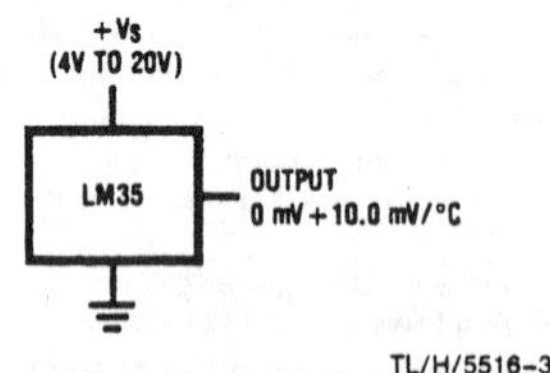

TL/H/5516-3

FIGURE 1. Basic Centigrade Temperature Sensor (+2°C to +150°C)

+Vs

LM35

VOUT

R1

−Vs

TL/H/5516-4

Choose $R_1 = -V_S/50\ \mu A$

V_{OUT} = +1,500 mV at +150°C
= +250 mV at +25°C
= −550 mV at −55°C

FIGURE 2. Full-Range Centigrade Temperature Sensor

Absolute Maximum Ratings (Note 10)

If Military/Aerospace specified devices are required, please contact the National Semiconductor Sales Office/Distributors for availability and specifications.

Supply Voltage	+35V to −0.2V
Output Voltage	+6V to −1.0V
Output Current	10 mA
Storage Temp., TO-46 Package,	−60°C to +180°C
TO-92 Package,	−60°C to +150°C
SO-8 Package,	−65°C to +150°C
TO-202 Package,	−65°C to +150°C
Lead Temp.:	
TO-46 Package, (Soldering, 10 seconds)	300°C
TO-92 Package, (Soldering, 10 seconds)	260°C
TO-202 Package, (Soldering, 10 seconds)	+230°C
SO Package (Note 12):	
Vapor Phase (60 seconds)	215°C
Infrared (15 seconds)	220°C
ESD Susceptibility (Note 11)	2500V
Specified Operating Temperature Range: T_{MIN} to T_{MAX} (Note 2)	
LM35, LM35A	−55°C to +150°C
LM35C, LM35CA	−40°C to +110°C
LM35D	0°C to +100°C

Electrical Characteristics (Note 1) (Note 6)

Parameter	Conditions	LM35A			LM35CA			Units (Max.)
		Typical	Tested Limit (Note 4)	Design Limit (Note 5)	Typical	Tested Limit (Note 4)	Design Limit (Note 5)	
Accuracy (Note 7)	$T_A = +25°C$	±0.2	±0.5		±0.2	±0.5		°C
	$T_A = -10°C$	±0.3			±0.3		±1.0	°C
	$T_A = T_{MAX}$	±0.4	±1.0		±0.4	±1.0		°C
	$T_A = T_{MIN}$	±0.4	±1.0		±0.4		±1.5	°C
Nonlinearity (Note 8)	$T_{MIN} \le T_A \le T_{MAX}$	**±0.18**		**±0.35**	**±0.15**		**±0.3**	°C
Sensor Gain (Average Slope)	$T_{MIN} \le T_A \le T_{MAX}$	**+10.0**	**+9.9, +10.1**		**+10.0**		**+9.9, +10.1**	mV/°C
Load Regulation (Note 3) $0 \le I_L \le 1$ mA	$T_A = +25°C$	±0.4	±1.0		±0.4	±1.0		mV/mA
	$T_{MIN} \le T_A \le T_{MAX}$	**±0.5**		**±3.0**	**±0.5**		**±3.0**	mV/mA
Line Regulation (Note 3)	$T_A = +25°C$	±0.01	±0.05		±0.01	±0.05		mV/V
	$4V \le V_S \le 30V$	**±0.02**		**±0.1**	**±0.02**		**±0.1**	mV/V
Quiescent Current (Note 9)	$V_S = +5V$, +25°C	56	67		56	67		µA
	$V_S = +5V$	**105**		**131**	**91**		**114**	µA
	$V_S = +30V$, +25°C	56.2	68		56.2	68		µA
	$V_S = +30V$	**105.5**		**133**	**91.5**		**116**	µA
Change of Quiescent Current (Note 3)	$4V \le V_S \le 30V$, +25°C	0.2	1.0		0.2	1.0		µA
	$4V \le V_S \le 30V$	**0.5**		**2.0**	**0.5**		**2.0**	µA
Temperature Coefficient of Quiescent Current		**+0.39**		**+0.5**	**+0.39**		**+0.5**	µA/°C
Minimum Temperature for Rated Accuracy	In circuit of *Figure 1*, $I_L = 0$	+1.5		+2.0	+1.5		+2.0	°C
Long Term Stability	$T_J = T_{MAX}$, for 1000 hours	±0.08			±0.08			°C

Note 1: Unless otherwise noted, these specifications apply: $-55°C \le T_J \le +150°C$ for the LM35 and LM35A; $-40° \le T_J \le +110°C$ for the LM35C and LM35CA; and $0° \le T_J \le +100°C$ for the LM35D. $V_S = +5Vdc$ and $I_{LOAD} = 50$ µA, in the circuit of *Figure 2.* These specifications also apply from +2°C to T_{MAX} in the circuit of *Figure 1.* Specifications in **boldface** apply over the full rated temperature range.

Note 2: Thermal resistance of the TO-46 package is 400°C/W, junction to ambient, and 24°C/W junction to case. Thermal resistance of the TO-92 package is 180°C/W junction to ambient. Thermal resistance of the small outline molded package is 220°C/W junction to ambient. Thermal resistance of the TO-202 package is 85°C/W junction to ambient. For additional thermal resistance information see table in the Applications section.

Electrical Characteristics (Note 1) (Note 6) (Continued)

Parameter	Conditions	LM35			LM35C, LM35D			Units (Max.)
		Typical	Tested Limit (Note 4)	Design Limit (Note 5)	Typical	Tested Limit (Note 4)	Design Limit (Note 5)	
Accuracy, LM35, LM35C (Note 7)	$T_A = +25°C$	±0.4	±1.0		±0.4	±1.0		°C
	$T_A = -10°C$	±0.5			±0.5		±1.5	°C
	$T_A = T_{MAX}$	±0.8	±1.5		±0.8		±1.5	°C
	$T_A = T_{MIN}$	±0.8		±1.5	±0.8		±2.0	°C
Accuracy, LM35D (Note 7)	$T_A = +25°C$				±0.6	±1.5		°C
	$T_A = T_{MAX}$				±0.9		±2.0	°C
	$T_A = T_{MIN}$				±0.9		±2.0	°C
Nonlinearity (Note 8)	$T_{MIN} \le T_A \le T_{MAX}$	**±0.3**		**±0.5**	**±0.2**		**±0.5**	°C
Sensor Gain (Average Slope)	$T_{MIN} \le T_A \le T_{MAX}$	**+10.0**	**+9.8, +10.2**		**+10.0**		**+9.8, +10.2**	mV/°C
Load Regulation (Note 3) $0 \le I_L \le 1$ mA	$T_A = +25°C$	±0.4	±2.0		±0.4	±2.0		mV/mA
	$T_{MIN} \le T_A \le T_{MAX}$	**±0.5**		**±5.0**	**±0.5**		**±5.0**	mV/mA
Line Regulation (Note 3)	$T_A = +25°C$	±0.01	±0.1		±0.01	±0.1		mV/V
	$4V \le V_S \le 30V$	**±0.02**		**±0.2**	**±0.02**		**±0.2**	mV/V
Quiescent Current (Note 9)	$V_S = +5V, +25°C$	56	80		56	80		μA
	$V_S = +5V$	**105**		**158**	**91**		**138**	μA
	$V_S = +30V, +25°C$	56.2	82		56.2	82		μA
	$V_S = +30V$	**105.5**		**161**	**91.5**		**141**	μA
Change of Quiescent Current (Note 3)	$4V \le V_S \le 30V, +25°C$	0.2	2.0		0.2	2.0		μA
	$4V \le V_S \le 30V$	**0.5**		**3.0**	**0.5**		**3.0**	μA
Temperature Coefficient of Quiescent Current		**+0.39**		**+0.7**	**+0.39**		**+0.7**	μA/°C
Minimum Temperature for Rated Accuracy	In circuit of *Figure 1*, $I_L = 0$	+1.5		+2.0	+1.5		+2.0	°C
Long Term Stability	$T_J = T_{MAX}$, for 1000 hours	±0.08			±0.08			°C

Note 3: Regulation is measured at constant junction temperature, using pulse testing with a low duty cycle. Changes in output due to heating effects can be computed by multiplying the internal dissipation by the thermal resistance.

Note 4: Tested Limits are guaranteed and 100% tested in production.

Note 5: Design Limits are guaranteed (but not 100% production tested) over the indicated temperature and supply voltage ranges. These limits are not used to calculate outgoing quality levels.

Note 6: Specifications in **boldface** apply over the full rated temperature range.

Note 7: Accuracy is defined as the error between the output voltage and 10mv/°C times the device's case temperature, at specified conditions of voltage, current, and temperature (expressed in °C).

Note 8: Nonlinearity is defined as the deviation of the output-voltage-versus-temperature curve from the best-fit straight line, over the device's rated temperature range.

Note 9: Quiescent current is defined in the circuit of *Figure 1*.

Note 10: Absolute Maximum Ratings indicate limits beyond which damage to the device may occur. DC and AC electrical specifications do not apply when operating the device beyond its rated operating conditions. See Note 1.

Note 11: Human body model, 100 pF discharged through a 1.5 kΩ resistor.

Note 12: See AN-450 "Surface Mounting Methods and Their Effect on Product Reliability" or the section titled "Surface Mount" found in a current National Semiconductor Linear Data Book for other methods of soldering surface mount devices.

National Semiconductor

LM135/LM235/LM335, LM135A/LM235A/LM335A Precision Temperature Sensors

General Description

The LM135 series are precision, easily-calibrated, integrated circuit temperature sensors. Operating as a 2-terminal zener, the LM135 has a breakdown voltage directly proportional to absolute temperature at +10 mV/°K. With less than 1Ω dynamic impedance the device operates over a current range of 400 μA to 5 mA with virtually no change in performance. When calibrated at 25°C the LM135 has typically less than 1°C error over a 100°C temperature range. Unlike other sensors the LM135 has a linear output.

Applications for the LM135 include almost any type of temperature sensing over a −55°C to +150°C temperature range. The low impedance and linear output make interfacing to readout or control circuitry especially easy.

The LM135 operates over a −55°C to +150°C temperature range while the LM235 operates over a −40°C to +125°C temperature range. The LM335 operates from −40°C to +100°C. The LM135/LM235/LM335 are available packaged in hermetic TO-46 transistor packages while the LM335 is also available in plastic TO-92 packages.

Features

- Directly calibrated in °Kelvin
- 1°C initial accuracy available
- Operates from 400 μA to 5 mA
- Less than 1Ω dynamic impedance
- Easily calibrated
- Wide operating temperature range
- 200°C overrange
- Low cost

Schematic Diagram

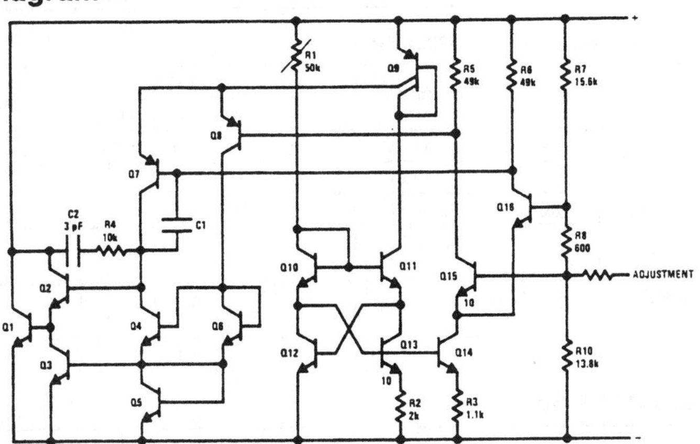

TL/H/5698-1

Connection Diagrams

TO-92 Plastic Package

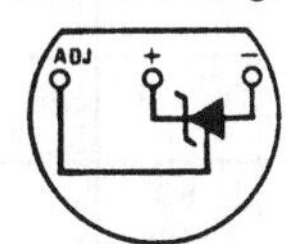

TL/H/5698-8

Bottom View

Order Number LM335Z or LM335AZ See NS Package Number Z03A

SO-8 Surface Mount Package

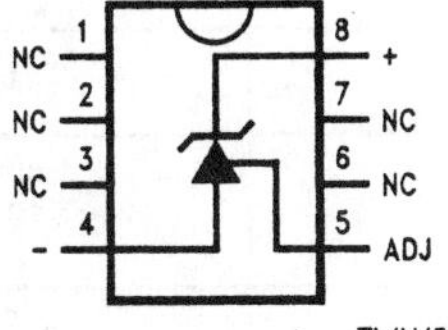

TL/H/5698-25

Order Number LM335M or LM335AM See NS Package Number M08A

TO-46 Metal Can Package*

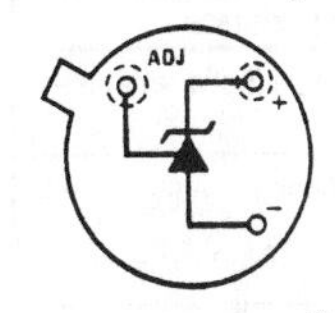

TL/H/5698-26

Bottom View

*Case is connected to negative pin

Order Number LM135H, LM135H-MIL, LM235H, LM335H, LM135AH, LM235AH or LM335AH See NS Package Number H03H

Absolute Maximum Ratings

If Military/Aerospace specified devices are required, please contact the National Semiconductor Sales Office/Distributors for availability and specifications. (Note 4)

Reverse Current 15 mA

Forward Current 10 mA

Storage Temperature
- TO-46 Package −60°C to +180°C
- TO-92 Package −60°C to +150°C
- SO-8 Package −65°C to +150°C

Specified Operating Temp. Range

	Continuous	Intermittent (Note 2)
LM135, LM135A	−55°C to +150°C	150°C to 200°C
LM235, LM235A	−40°C to +125°C	125°C to 150°C
LM335, LM335A	−40°C to +100°C	100°C to 125°C

Lead Temp. (Soldering, 10 seconds)
- TO-92 Package: 260°C
- TO-46 Package: 300°C
- SO-8 Package: 300°C
  - Vapor Phase (60 seconds) 215°C
  - Infrared (15 seconds) 220°C

Temperature Accuracy LM135/LM235, LM135A/LM235A (Note 1)

Parameter	Conditions	LM135A/LM235A			LM135/LM235			Units
		Min	Typ	Max	Min	Typ	Max	
Operating Output Voltage	$T_C = 25°C$, $I_R = 1$ mA	2.97	2.98	2.99	2.95	2.98	3.01	V
Uncalibrated Temperature Error	$T_C = 25°C$, $I_R = 1$ mA		0.5	1		1	3	°C
Uncalibrated Temperature Error	$T_{MIN} \le T_C \le T_{MAX}$, $I_R = 1$ mA		1.3	2.7		2	5	°C
Temperature Error with 25°C Calibration	$T_{MIN} \le T_C \le T_{MAX}$, $I_R = 1$ mA		0.3	1		0.5	1.5	°C
Calibrated Error at Extended Temperatures	$T_C = T_{MAX}$ (Intermittent)		2			2		°C
Non-Linearity	$I_R = 1$ mA		0.3	0.5		0.3	1	°C

Temperature Accuracy LM335, LM335A (Note 1)

Parameter	Conditions	LM335A			LM335			Units
		Min	Typ	Max	Min	Typ	Max	
Operating Output Voltage	$T_C = 25°C$, $I_R = 1$ mA	2.95	2.98	3.01	2.92	2.98	3.04	V
Uncalibrated Temperature Error	$T_C = 25°C$, $I_R = 1$ mA		1	3		2	6	°C
Uncalibrated Temperature Error	$T_{MIN} \le T_C \le T_{MAX}$, $I_R = 1$ mA		2	5		4	9	°C
Temperature Error with 25°C Calibration	$T_{MIN} \le T_C \le T_{MAX}$, $I_R = 1$ mA		0.5	1		1	2	°C
Calibrated Error at Extended Temperatures	$T_C = T_{MAX}$ (Intermittent)		2			2		°C
Non-Linearity	$I_R = 1$ mA		0.3	1.5		0.3	1.5	°C

Electrical Characteristics (Note 1)

Parameter	Conditions	LM135/LM235 LM135A/LM235A			LM335 LM335A			Units
		Min	Typ	Max	Min	Typ	Max	
Operating Output Voltage Change with Current	400 μA $\le I_R \le$ 5 mA At Constant Temperature		2.5	10		3	14	mV
Dynamic Impedance	$I_R = 1$ mA		0.5			0.6		Ω
Output Voltage Temperature Coefficient			+10			+10		mV/°C
Time Constant	Still Air 100 ft/Min Air Stirred Oil		80 10 1			80 10 1		sec sec sec
Time Stability	$T_C = 125°C$		0.2			0.2		°C/khr

Note 1: Accuracy measurements are made in a well-stirred oil bath. For other conditions, self heating must be considered.

Note 2: Continuous operation at these temperatures for 10,000 hours for H package and 5,000 hours for Z package may decrease life expectancy of the device.

Note 3: Thermal Resistance

	TO-92	TO-46	SO-8
θ_{JA} (junction to ambient)	202°C/W	400°C/W	165°C/W
θ_{JC} (junction to case)	170°C/W	N/A	N/A

Note 4: Refer to RETS135H for military specifications.

Reprinted with permission of National Semiconductor Corporation

LM135/LM235/LM335, LM135A/LM235A/LM335A

Typical Performance Characteristics

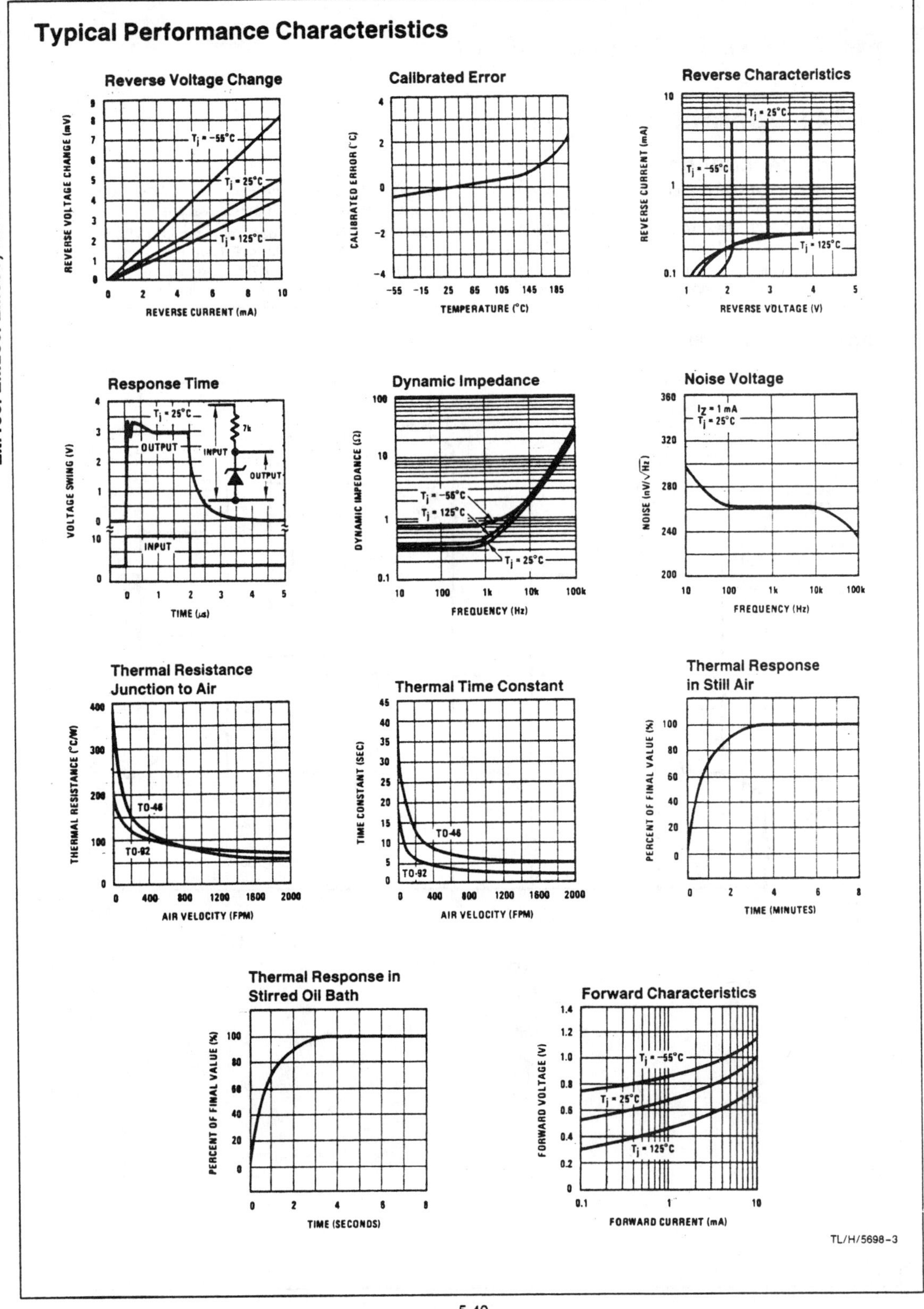

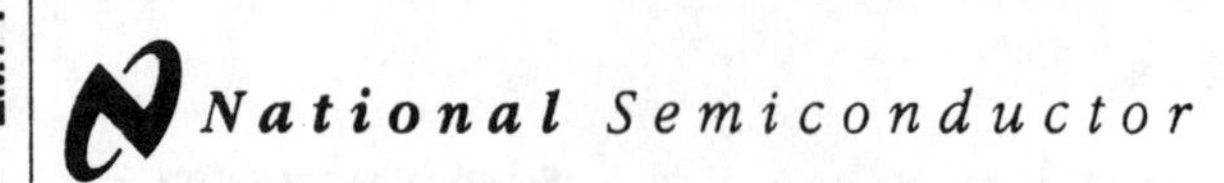

LM741 Operational Amplifier

General Description

The LM741 series are general purpose operational amplifiers which feature improved performance over industry standards like the LM709. They are direct, plug-in replacements for the 709C, LM201, MC1439 and 748 in most applications.

The amplifiers offer many features which make their application nearly foolproof: overload protection on the input and output, no latch-up when the common mode range is exceeded, as well as freedom from oscillations.

The LM741C/LM741E are identical to the LM741/LM741A except that the LM741C/LM741E have their performance guaranteed over a 0°C to +70°C temperature range, instead of −55°C to +125°C.

Schematic Diagram

NON-INVERTING INPUT 3, 2 INVERTING INPUT, Q1, Q2, Q3, Q4, Q5, Q6, Q7, Q8, Q9, Q10, Q11, Q12, Q13, Q14, Q15, Q15, Q17, Q20, Q22, R1 1 K, R2 1 K, R3 50 K, R4 5 K, R5 39 K, R7 4.5 K, R8 7.5 K, R9 25, R10 50, R11 50, R12 50 K, C1 30 pF, OFFSET NULL 1, 5 OFFSET NULL, 7 V+, 6 OUTPUT, 4 V−

TL/H/9341-1

Offset Nulling Circuit

−, +, LM741, OUTPUT, 10 kΩ, V−

TL/H/9341-7

Absolute Maximum Ratings

If Military/Aerospace specified devices are required, please contact the National Semiconductor Sales Office/ Distributors for availability and specifications.

(Note 5)

	LM741A	LM741E	LM741	LM741C
Supply Voltage	±22V	±22V	±22V	±18V
Power Dissipation (Note 1)	500 mW	500 mW	500 mW	500 mW
Differential Input Voltage	±30V	±30V	±30V	±30V
Input Voltage (Note 2)	±15V	±15V	±15V	±15V
Output Short Circuit Duration	Continuous	Continuous	Continuous	Continuous
Operating Temperature Range	−55°C to +125°C	0°C to +70°C	−55°C to +125°C	0°C to +70°C
Storage Temperature Range	−65°C to +150°C	−65°C to +150°C	−65°C to +150°C	−65°C to +150°C
Junction Temperature	150°C	100°C	150°C	100°C
Soldering Information				
N-Package (10 seconds)	260°C	260°C	260°C	260°C
J- or H-Package (10 seconds)	300°C	300°C	300°C	300°C
M-Package				
Vapor Phase (60 seconds)	215°C	215°C	215°C	215°C
Infrared (15 seconds)	215°C	215°C	215°C	215°C

See AN-450 "Surface Mounting Methods and Their Effect on Product Reliability" for other methods of soldering surface mount devices.

	LM741A	LM741E	LM741	LM741C
ESD Tolerance (Note 6)	400V	400V	400V	400V

Electrical Characteristics (Note 3)

Parameter	Conditions	LM741A/LM741E			LM741			LM741C			Units
		Min	Typ	Max	Min	Typ	Max	Min	Typ	Max	
Input Offset Voltage	$T_A = 25°C$										
	$R_S \leq 10\ k\Omega$					1.0	5.0		2.0	6.0	mV
	$R_S \leq 50\Omega$		0.8	3.0							mV
	$T_{AMIN} \leq T_A \leq T_{AMAX}$										
	$R_S \leq 50\Omega$			4.0							mV
	$R_S \leq 10\ k\Omega$						6.0			7.5	mV
Average Input Offset Voltage Drift				15							μV/°C
Input Offset Voltage Adjustment Range	$T_A = 25°C, V_S = \pm 20V$	±10				±15			±15		mV
Input Offset Current	$T_A = 25°C$		3.0	30		20	200		20	200	nA
	$T_{AMIN} \leq T_A \leq T_{AMAX}$			70		85	500			300	nA
Average Input Offset Current Drift				0.5							nA/°C
Input Bias Current	$T_A = 25°C$		30	80		80	500		80	500	nA
	$T_{AMIN} \leq T_A \leq T_{AMAX}$			0.210			1.5			0.8	μA
Input Resistance	$T_A = 25°C, V_S = \pm 20V$	1.0	6.0		0.3	2.0		0.3	2.0		MΩ
	$T_{AMIN} \leq T_A \leq T_{AMAX}$, $V_S = \pm 20V$	0.5									MΩ
Input Voltage Range	$T_A = 25°C$							±12	±13		V
	$T_{AMIN} \leq T_A \leq T_{AMAX}$				±12	±13					V
Large Signal Voltage Gain	$T_A = 25°C, R_L \geq 2\ k\Omega$										
	$V_S = \pm 20V, V_O = \pm 15V$	50									V/mV
	$V_S = \pm 15V, V_O = \pm 10V$				50	200		20	200		V/mV
	$T_{AMIN} \leq T_A \leq T_{AMAX}$, $R_L \geq 2\ k\Omega$,										
	$V_S = \pm 20V, V_O = \pm 15V$	32									V/mV
	$V_S = \pm 15V, V_O = \pm 10V$				25			15			V/mV
	$V_S = \pm 5V, V_O = \pm 2V$	10									V/mV

Electrical Characteristics (Note 3) (Continued)

Parameter	Conditions	LM741A/LM741E			LM741			LM741C			Units
		Min	Typ	Max	Min	Typ	Max	Min	Typ	Max	
Output Voltage Swing	$V_S = \pm 20V$										
	$R_L \geq 10\ k\Omega$	±16									V
	$R_L \geq 2\ k\Omega$	±15									V
	$V_S = \pm 15V$										
	$R_L \geq 10\ k\Omega$				±12	±14		±12	±14		V
	$R_L \geq 2\ k\Omega$				±10	±13		±10	±13		V
Output Short Circuit Current	$T_A = 25°C$	10	25	35		25			25		mA
	$T_{AMIN} \leq T_A \leq T_{AMAX}$	10		40							mA
Common-Mode Rejection Ratio	$T_{AMIN} \leq T_A \leq T_{AMAX}$										
	$R_S \leq 10\ k\Omega$, $V_{CM} = \pm 12V$				70	90		70	90		dB
	$R_S \leq 50\Omega$, $V_{CM} = \pm 12V$	80	95								dB
Supply Voltage Rejection Ratio	$T_{AMIN} \leq T_A \leq T_{AMAX}$, $V_S = \pm 20V$ to $V_S = \pm 5V$										
	$R_S \leq 50\Omega$	86	96								dB
	$R_S \leq 10\ k\Omega$				77	96		77	96		dB
Transient Response	$T_A = 25°C$, Unity Gain										
Rise Time			0.25	0.8		0.3			0.3		μs
Overshoot			6.0	20		5			5		%
Bandwidth (Note 4)	$T_A = 25°C$	0.437	1.5								MHz
Slew Rate	$T_A = 25°C$, Unity Gain	0.3	0.7			0.5			0.5		V/μs
Supply Current	$T_A = 25°C$					1.7	2.8		1.7	2.8	mA
Power Consumption	$T_A = 25°C$										
	$V_S = \pm 20V$		80	150							mW
	$V_S = \pm 15V$					50	85		50	85	mW
LM741A	$V_S = \pm 20V$										
	$T_A = T_{AMIN}$			165							mW
	$T_A = T_{AMAX}$			135							mW
LM741E	$V_S = \pm 20V$										
	$T_A = T_{AMIN}$			150							mW
	$T_A = T_{AMAX}$			150							mW
LM741	$V_S = \pm 15V$										
	$T_A = T_{AMIN}$					60	100				mW
	$T_A = T_{AMAX}$					45	75				mW

Note 1: For operation at elevated temperatures, these devices must be derated based on thermal resistance, and T_j max. (listed under "Absolute Maximum Ratings"). $T_j = T_A + (\theta_{jA} P_D)$.

Thermal Resistance	Cerdip (J)	DIP (N)	HO8 (H)	SO-8 (M)
θ_{jA} (Junction to Ambient)	100°C/W	100°C/W	170°C/W	195°C/W
θ_{jC} (Junction to Case)	N/A	N/A	25°C/W	N/A

Note 2: For supply voltages less than ±15V, the absolute maximum input voltage is equal to the supply voltage.

Note 3: Unless otherwise specified, these specifications apply for $V_S = \pm 15V$, $-55°C \leq T_A \leq +125°C$ (LM741/LM741A). For the LM741C/LM741E, these specifications are limited to $0°C \leq T_A \leq +70°C$.

Note 4: Calculated value from: BW (MHz) = 0.35/Rise Time(μs).

Note 5: For military specifications see RETS741X for LM741 and RETS741AX for LM741A.

Note 6: Human body model, 1.5 kΩ in series with 100 pF.

SPR-1-08/SPR-1-10 Silicon Multi-Channel Detector Array

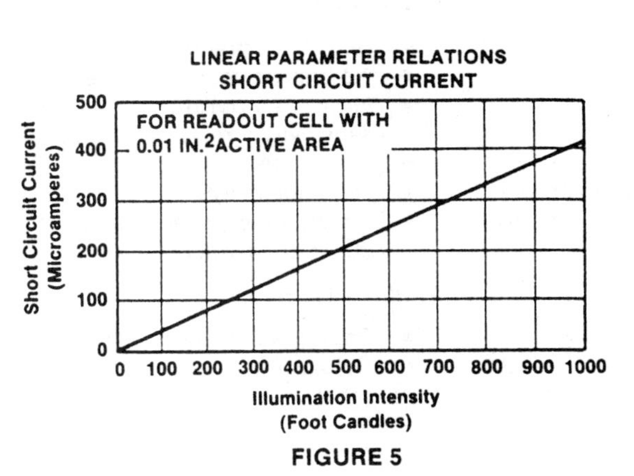

FIGURE 5

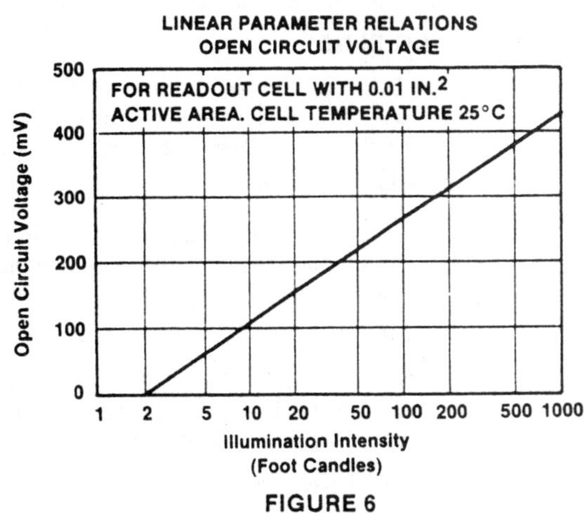

FIGURE 6

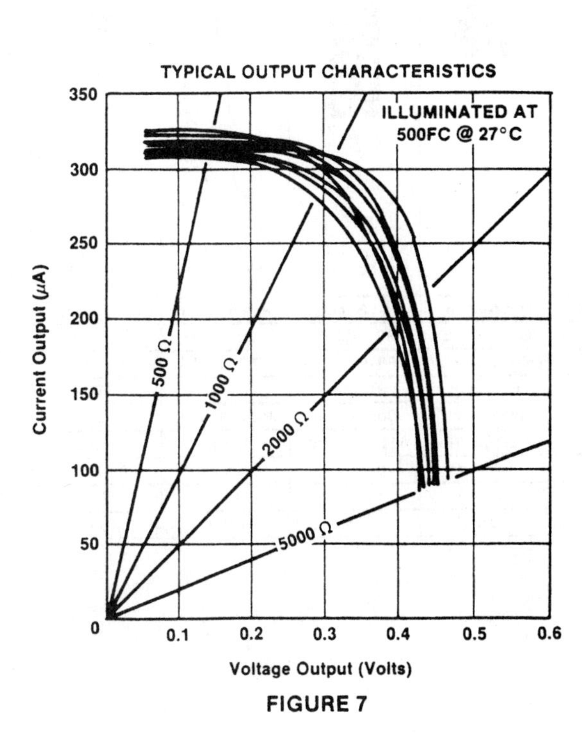

FIGURE 7

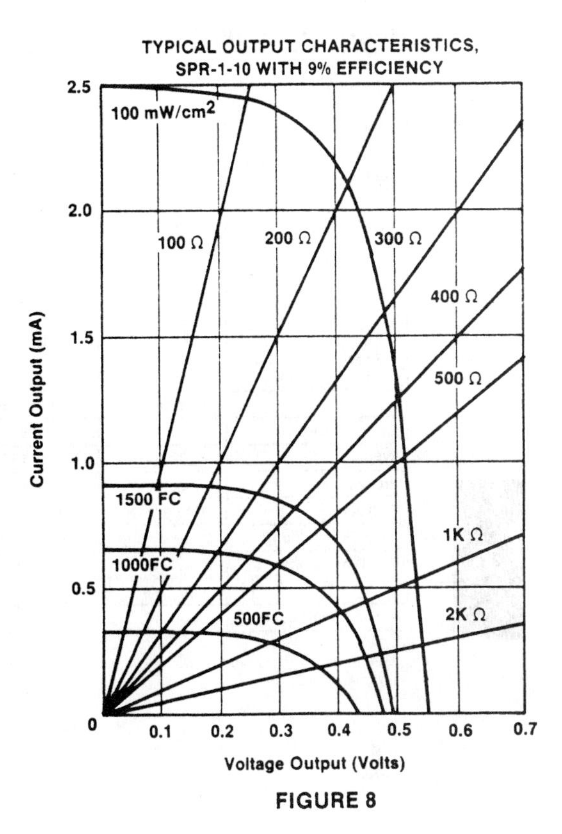

FIGURE 8

TOI Texas Optoelectronics, Inc.

Spectra-Band Photocell Series

A series of spectral-response silicon photocells designed for unique product applications.

VIO-BLUE

Enhanced violet and blue response. Also can be used in U.V. detection because of high sensitivity to short wavelength radiation.

GREEN BLAZE

Photopic curve response for use in innumerable light response applications — with high reliability and low cost.

INFRA-R

Visible cut-off, high infrared response. Solves ambient light problems in IR activated photoelectric applications.

FEATURES

- Select spectral response
- No bias power source needed
- High temperature stability and high sensitivity through silicon construction
- Low noise
- High reliability
- A wide variety of sizes and packages, special geometries available

APPLICATIONS

- Photographic equipment
- Color, pattern recognition equipment
- Light discriminating systems

SPECTRA-BAND PHOTOCELLS

TOIs' special spectral response photocells are designed for the photographic industry, photometric instrumentation, and photoelectric control/switching applications.

MECHANICAL SPECIFICATIONS

Spectra-Band Cell Configurations					Modified TO-206 AA (TO-18)	TO-205 AA (TO-5)	TO-233 AA (TO-8)
	Part Number	Part Number	Part Number	Part Number	Part Number	Part Number	Part Number
GREEN BLAZE	GB02505EPL	GB0505EPL	GB1010EPL	GB1020EPL	GBTO-18	GBTO-5	GBTO-8
INFRA-R	FR02505EPL	FR0505EPL	*FR1010EPL	FR1020EPL	FRTO-18	FRTO-5	FRTO-8
VIO-BLUE	VB02505EPL	VB0505EPL	VB1010EPL	VB1020EPL	VBTO-18	VBTO-5	VBTO-8
Package	Coated Cell	Coated Cell	Coated Cell	Coated Cell	Modified TO-18	TO-5	TO-8
Lead Termination	6" Length Std.	6" Length Std.	6" Length Std.	6" Length Std.	Leads	Leads	Leads
Cell Dimensions In.	0.1 x 0.2	0.2 x 0.2	0.4 x 0.4	0.4 x 0.8	0.055 x 0.055	0.1 x 0.2	0.28 x 0.28
Cm.	0.25 x 0.5	0.5 x 0.5	1.0 x 1.0	1.0 x 2.0	0.14 x 0.14	0.25 x 0.5	0.72 x 0.72
Active Area (Sq. Cm.)	0.1	0.2	0.9	1.8	0.018	0.1	0.5

714 Shepherd Drive · Garland, Texas 75042 · 214/487-0085

Spectra-Band Photocell Series

TYPICAL PERFORMANCE CHARACTERISTICS (Continued)

INFRA-R SERIES

Parameter	Symbol	Unit	Test Condition	FR02505EPL	FR0505EPL	FR1010EPL	FR1020EPL	FRTO-18	FRTO-5	FRTO-8
Short Circuit Current	I_{SC}	mA	100 mW/cm^2, AM1 Solar Radiation	1.3	2.6	11.5	23.0	0.3	1.3	6.4
Open Circuit Voltage	V_{OC}	Volts	100 mW/cm^2, AM1 Solar Radiation	0.55	0.55	0.55	0.55	0.55	0.55	0.55
Forward Voltage	V_F	Volts	I_F = 1 mA	0.50	0.50	0.45	0.40	0.50	0.50	0.45
Dark Current	I_D	μA	V_R = 0.1 V	0.2	0.4	0.8	0.9	0.2	0.2	0.5
Capacitance	C_T	pF	V_R = 0 V	1.0	3.0	10.0	15.0	1.0	1.0	8.0
Responsivity	R_e	A/W	λp = 900 nm, V_R = 0	0.45	0.45	0.45	0.45	0.40	0.40	0.40

VIO-BLUE SERIES

Parameter	Symbol	Unit	Test Condition	VB02505EPL	VB0505EPL	VB1010EPL	VB1020EPL	VBTO-18	VBTO-5	VBTO-8
Short Circuit Current	I_{SC}	mA	100 mW/cm^2, AM1 Solar Radiation	2.3	4.7	21.0	42.0	0.5	2.3	11.6
Open Circuit Voltage	V_{OC}	Volts	100 mW/cm^2, AM1 Solar Radiation	0.55	0.55	0.55	0.55	0.55	0.55	0.55
Forward Voltage	V_F	Volts	I_F = 1 mA	0.50	0.50	0.45	0.40	0.50	0.50	0.45
Dark Current	I_D	μA	V_R = 0.1 V	0.2	0.4	0.8	0.9	0.2	0.2	0.5
Capacitance	C_T	pF	V_R = 0 V	1.0	3.0	10.0	15.0	1.0	1.0	8.0
Responsivity	R_e	A/W	λp = 900 nm, V_R = 0	0.48	0.48	0.48	0.48	0.44	0.44	0.44

GREEN-BLAZE SERIES

Parameter	Symbol	Unit	Test Condition	GB02505EPL	GB0505EPL	GB1010EPL	GB1020EPL	GBTO-18	GBTO-5	GBTO-8
Short Circuit Current	I_{SC}	mA	100 mW/cm^2, AM1 Solar Radiation	0.27	0.55	2.5	5.0	0.06	0.27	1.38
Open Circuit Voltage	V_{OC}	Volts	100 mW/cm^2, AM1 Solar Radiation	0.47	0.47	0.47	0.47	0.47	0.47	0.47
Forward Voltage	V_F	Volts	I_F = 1 mA	0.50	0.45	0.45	0.40	0.50	0.50	0.45
Dark Current	I_D	μA	V_R = 0.1 V	0.3	0.4	0.8	1.0	0.3	0.3	0.5
Capacitance	C_T	pF	V_R = 0 V	1.0	2.0	3.0	5.0	1.0	1.0	2.0
Responsivity	R_e	A/W	λp = 555 nm	0.20	0.20	0.20	0.20	0.19	0.19	0.19

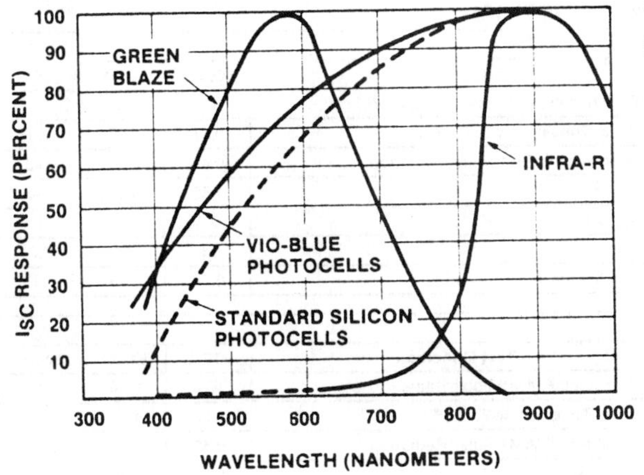

TYPICAL SPECTRAL RESPONSE CHARACTERISTICS — NORMALIZED

TYPICAL SHORT CIRCUIT CURRENT (I_{SC}) RESPONSE

- Standard Silicon Photovoltaic Cell (at 900 nm) ~ 0.48 A/W
- Vio-Blue (at 900 nm) ~ 0.48 A/W
- Green Blaze (at 555 nm) ~ 0.20 A/W
- Infra-R (at 900 nm) ~ 0.45 A/W

714 Shepherd Drive · Garland, Texas 75042 · 214/487-0085

TOI Texas Optoelectronics, Inc.

Silicon Photocell Sensors

SILICON PHOTOCELL SENSORS

TOI silicon photocells are employed in photometer, switching, position detection, tape and disc EOT-BOT sensing, solar energy conversion, and other numerous applications. Silicon photosensors with special geometries, spectral response and switching characteristics, are available on a custom basis, and are widely used in the optical encoder, character recognition, and optical instrumentation fields.

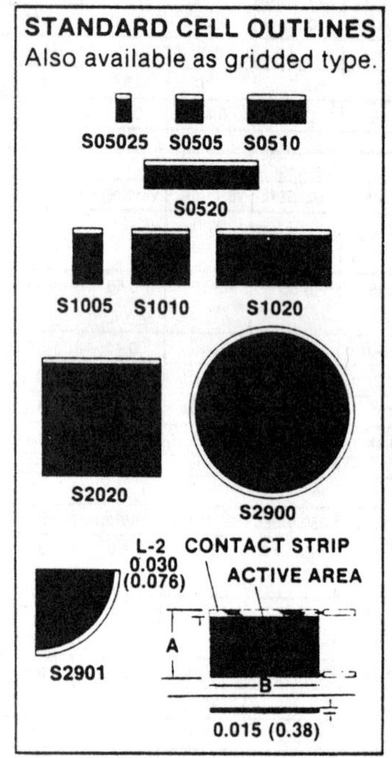

Standard Size Part Numbers	Cell Dimensions		Photo Active Area		(1) Test Voltage (Volts)
	in.	cm.	in.^{2}	cm^2	
S05025	0.20 x 0.10	0.5 x 0.25	.017	0.1	.43
S0505	0.20 x 0.20	0.5 x 0.5	.034	0.2	.43
S0510	0.20 x 0.40	0.5 x 1.0	.068	0.4	.43
S0520	0.20 x 0.80	0.5 x 2.0	.136	0.8	.43
S1005	0.40 x 0.20	1.0 x 0.5	.074	0.4	.43
S1010	0.40 x 0.40	1.0 x 1.0	.148	0.9	.43
S1020	0.40 x 0.80	1.0 x 2.0	.296	1.9	.43
S2020	0.80 x 0.80	2.0 x 2.0	.620	3.8	.43
S2900	1.125 Dia.	2.86	.88	5.7	.43
S2901	Quarter Section of S2900	—	.22	1.4	.43

NOTE: (1) Irradiance: 100 mW/cm^2, AM1 solar radiation.

Part Number Code for Ordering Silicon Light Sensors

EXAMPLE: S 05 05 G E 6 PL

Silicon	"A" Width	"B" Length	Gridded Type	Device Type	Minimum Conversion Efficiency	Leads If Desired
(Outline L-2)	05 = 0.20" (0.5 cm) 10 = 0.40" (1.0 cm) 20 = 0.80" (2.0 cm)		Add "G" for cells 0.4" x 0.4" (1.0 x 1.0 cm) and larger	"E" P on N	5% to 10% (6 = 6%, etc.)	PL — (Pigtail Leads)

TYPICAL PERFORMANCE CHARACTERISTICS

STANDARD SILICON PHOTOCELL

Parameter	Symbol	Unit	Test Condition	S05025	S0505	S0510	S0520	S1005
Short Circuit Current	I_{SC}	mA	100 mW/cm^2, AM1 Solar Radiation	1.8	3.8	7.5	15.0	7.5
Short Circuit Current	I_{SC}	mA	100 fc, Tungsten @ 2870°K	0.07	0.13	0.27	0.54	0.27
Open Circuit Voltage	V_{OC}	Volts	100 mW/cm^2, AM1 Solar Radiation	0.43	0.43	0.43	0.43	0.43
Forward Voltage	V_F	Volts	I_F = 1 mA	0.50	0.50	0.42	0.42	0.42
Dark Current	I_D	μA	V_R = 0.1 V	0.3	0.5	0.6	0.8	0.6
Capacitance	C_T	pF	V_R = 0 V	1.5	2.4	5.0	10.0	5.0
Responsivity	R_e	A/W	λp = 900 nm	0.48	0.48	0.48	0.48	0.48

STANDARD SILICON PHOTOCELL (Continued)

Parameter	Symbol	Unit	Test Condition	S1010	S1020	S2020	S2900	S2901
Short Circuit Current	I_{SC}	mA	100 mW/cm^2, AM1 Solar Radiation	17.0	35.0	72.0	105.0	27.0
Short Circuit Current	I_{SC}	mA	100 fc, Tungsten @ 2870°K	0.55	1.10	2.2	3.3	0.8
Open Circuit Voltage	V_{OC}	Volts	100 mW/cm^2, AM1 Solar Radiation	0.43	0.43	0.43	0.43	0.43
Forward Voltage	V_F	Volts	I_F = 1 mA	0.42	0.40	0.30	0.25	0.40
Dark Current	I_D	μA	V_R = 0.1 V	0.8	1.8	25.0	100.0	100.0
Capacitance	C_T	pF	V_R = 0 V	20.0	25.0	70.0	90.0	25.0
Responsivity	R_e	A/W	λp = 900 nm	0.48	0.48	0.48	0.48	0.48

714 Shepherd Drive · Garland, Texas 75042 · 214/487-0085

APPENDIX B

Tutorial on Using Electronics Workbench

You must have version 5.1 of Electronics Workbench (EWB) loaded on your computer prior to performing the exercise below. Once you have brought the program up with your windows "Start" command, then examine the top of the screen. At the top left of the screen you will see the six menus available to you. They are "*File*, *Edit*, *Circuit*, *Analysis*, *Window*, and *Help*." By clicking on each of these with your left mouse button you will see the pull-down commands available to you. Since this is a tutorial you will not have an opportunity to use many of the 40 commands.

Below the menus you will see the circuit tool bar. These represent commands to do things to the circuit you will construct below on the breadboard space. The next row contains the electronic parts bins. As you can see by clicking on the first parts bin (battery symbol) there are quite a large number of sources available for your use. There are 13 parts bins, and there is a special one called subcircuits on the far left in which you can store circuits as a module. The ON/OFF switch and Pause/Resume button is on the top right of the screen. Scroll bars are located on the side and bottom of the breadboard window for moving your circuits around on the breadboard. Now, let's create a circuit on the white breadboard area.

Let's start with a circuit you are familiar with, a simple amplifier using discrete components. The circuit is shown below.

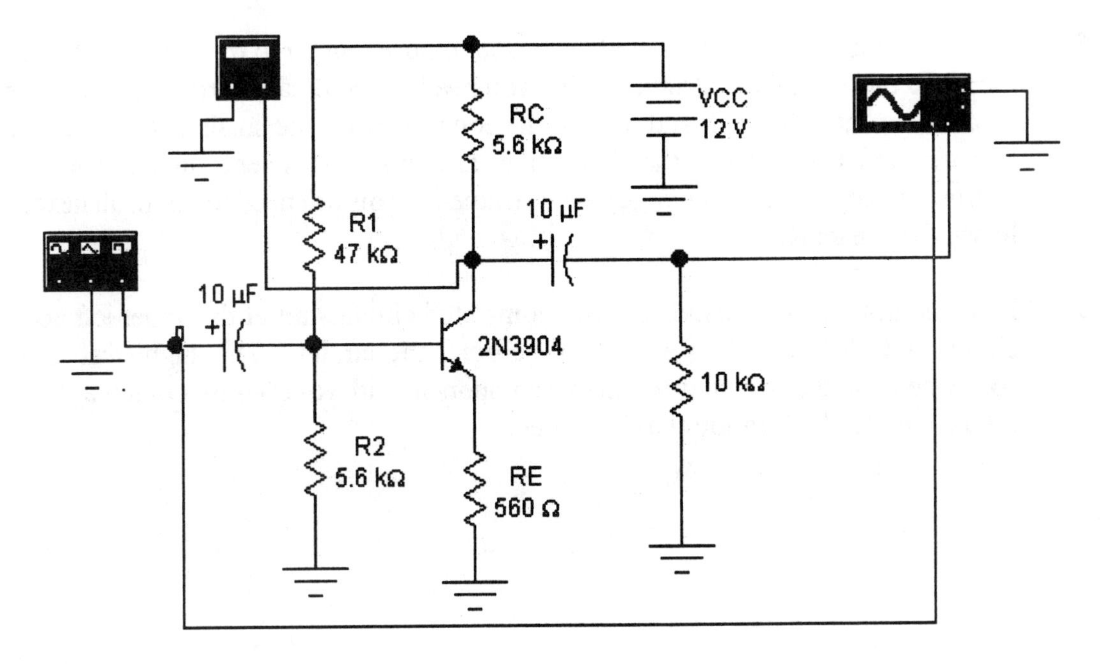

Figure Appendix-B

Step 1. Open the resistor parts bin (click with left mouse button, LMB) and then click and drag the resistors out onto the breadboard area one at a time. The resistors and other components can be rotated by using the rotate icon in the circuits tool bar or by using the right mouse button and selecting rotate. Do this.

Step 2. Open the transistor parts bin and drag out an NPN transistor. You may select the transistor type by double-clicking the LMB and then select. Select either the 2N3904 or one from a manufacturer part assortment that is listed. You may also want to bring out your power supply source by clicking on the battery symbol. This bin also contains the ground symbol, so drag the number of these that you need out also.

Step 3. Assemble your circuit by placing (click and drag to position) the transistor somewhere near the middle and then placing the resistors in their respective positions. Make sure you leave adequate room between parts so that you may make connections with the instruments. Place the mouse pointer near the end of the component, then click and drag a wire from that component to the end of the next part. Release gently so that the wire is not jerked away. Continue doing this until your circuit is completed. Note that you have electrolytic and non-electrolytic capacitors.

Step 4. The components now can be **labeled** and **values** placed on them. To do this double-click on the component. This will bring up instructions (which look like file folders) so that you can label the components and place values on them. Note that you have up and down arrows on the values to set common electronic metric units on the parts.

Step 5. Now that your circuit is completed move your mouse pointer to the instrument bin. Click the LMB and drag the instruments that you need to the breadboard space. Instruments can be connected by click and drag at the connectors on the small instrument icons. If you are confused by where the connections are on an instrument, then double-click the instrument icon to zoom (enlarge) the instrument. You will need to zoom all instruments to see their readings.

Step 6. If you do not like the position of your completed circuit start at the upper left above the circuit and click and drag a rectangle around the circuit. This highlights all circuit components so that you may touch one component with your mouse pointer and drag the total circuit to the location that is desired.

Step 7. Now that the circuit is completed, set the controls on all of your instruments. Set the function generator to 100 mV of sine wave at a frequency of 1 kHz. The function generator readout is peak units. Set the digital multimeter (DMM) to dc volts (flat line for dc and V for volts). Set the oscilloscope controls to 50 mV per division on channel A and 200 mV per division on channel B. Make sure that CH A is connected to the input and CH B is connected to the load.

Step 8. You may turn your circuit on by pointing the mouse at the ON/OFF switch in the upper right-hand corner of the screen. Make sure your instruments are zoomed open so that you can see the readings. You may pause the circuit operation by clicking the mouse pointer at the pause button below the ON/OFF switch.

Step 9. Go to the Analysis menu at the top of the screen and set the instruments to pause after each screen. The steps are a) point and click *Analysis* menu, b) click on analysis options, c) click on instruments and d) click on box to pause after each screen.

Step 10. You can save the circuit to your files disk by going to the *File* menu and clicking on save. Make sure you are saving the file to your files disk and not to the hard drive of the computer. If you have a printer connected to the computer you can go to the *File* menu and click on print. Most printers can be used. If set up for the printer is needed go to the *File* menu to select printer.

APPENDIX C
Partial List of Vendors

The following is a list of vendors that are reliable sources of electronic components.

Digi-Key Corporation, 1-800-344-4539 when ordering parts
Can be contacted on the Web at www.digikey.com

Electronic School Supply, 1-800-726-0084
Can be contacted on the Web at essinc@lightspeed.net
They do a good job putting together parts kits for courses

Jameco Electronics, 1-800-831-4242
Can be contacted on the Web at www.jameco.com

Mouser Electronics, 1-800-346-6873 when ordering parts
Can be contacted on the Web at www.mouser.com

Electronix Express, 1-800-972-2225
Can be contacted on the Web at www.electron@elexp.com
They do parts kits for school orders.

Kelvin Electronics, 1-800-KELVIN9
Can be contacted on the Web at www.kelvin.com

There are many others that could be mentioned. If you want information on the Web, it is best to know the company name ahead of time, otherwise you will have thousands of pages to look through.

APPENDIX D

Parts List for Hardware Labs

<u>Integrated Circuits</u>

1 each of the following:

- LM741 op-amp
- LF351N bi-fet op-amp
- 7408 quad AND gate
- LM335 temperature sensor
- LM340 or 7805 five volt
- amp regulator
- LM380 power amp
- AD620 instrumentation amplifier
- DAC08 digital to analog converter
- ADC-0801 analog to digital converter

2 each of the following:

- LM311 precision comparator
- 555 timers

Note: If you wish to do other temperature sensor exercises 1-LM34 and/or LM35 are required.

<u>Transistors, SCR, etc.</u>

1 each of the following:

- SCR 106B or similar
- 2N3904
- 2N3906
- 2N4123
- 2N3053
- 2N2955
- 2N3055 transistors
- 2N5064 SCR

<u>Diodes</u>

1 each of the following:

- 1N914
- 1N4001
- Diode Bridge @ 1 amp

1 each of the following zener diodes:

- 1N4730
- 1N4733
- 1N4735

2 each of the following:

- 1N4148
- 1N4739 (zener diode)

<u>Resistors</u>

1 each of the following ¼-watt or ½-watt film resistors:

2.7 ohm	100 ohm	820 ohm	1.2 K-ohm
2.7 K-ohm	3.9 K-ohm	33 K-ohm	100 K-ohm
220 K-ohm	390 K-ohm	470 K-ohm	1 M-ohm
3.9 M-ohm.			

2 each of the following ¼-watt or ½-watt film resistors:

2.2 M-ohm	47 K-ohm	10 ohm	1 K-ohm
2.2 K-ohm			

3 each of the following ¼-watt or ½-watt film resistors:

15 K-ohm	120 ohm

8 each of the following ¼-watt or ½-watt film resistors:

330 ohm	10 K-ohm

Power Resistors
2 each: 0.47 ohm @ 2 watt
1 each: 10 ohm @ 10 watt

Potentiometers
1 each of the following:

1K	10K	20K	50K
100K	1M. ½-watt to ¾-watt single-turn or multi-turn if more accuracy is desired		

Capacitors
1 each of the following:

150 pf	0.47 uf	1 uf	10 uf
50 uf	100 uf	470 uf electrolytics of at least 35 WVDC.	

2 each: 0.1 uf

3 each: .01 uf

Miscellaneous Parts
1 each of the following:

miniature piezo audible buzzer	12 VDC relay, low current of 100 mA or less to activate,
12 volt, 5 watt lamp	12VCT @ 1 amp transformer
solar cell (see Appendix A)	high output infrared LED
miniature speaker	5W speaker
120 ohm wire strain gage	
microphone omnidirectional Electret condenser microphone cartridge	

8 each: red LEDs